RF & Microwave Multi-mode and Tunable Filters

Fushun Nian · Dinghong Jia · Xiaoguo Huang ·
Shengjun Wei

RF & Microwave Multi-mode and Tunable Filters

Springer

Fushun Nian
Ceyear Technology Co. Ltd.
Qingdao, Shandong, China

Dinghong Jia
Ceyear Technology Co. Ltd.
Qingdao, Shandong, China

Xiaoguo Huang
Jiangnan Electronics Communications
Research Institute
Jiaxing, Zhejiang, China

Shengjun Wei
College of Physics and Electronic
Engineering
Sichuan Normal University
Chengdu, Sichuan, China

ISBN 978-981-92-0576-9 ISBN 978-981-92-0577-6 (eBook)
https://doi.org/10.1007/978-981-92-0577-6

This book is funded by the National Natural Science Foundation project (U24B6013).

This Springer imprint is published by the registered company Springer Nature Singapore Pte Ltd.
The registered company address is: 152 Beach Road, #21-01/04 Gateway East, Singapore 189721, Singapore

Preface

With the emerging multi-scenario applications such as 5G communication, radar, the Internet of Things (IoT), and satellite internet putting forward stringent requirements for signal filtering technology, including multimode adaptability, high integration density, and low insertion loss, multimode and tunable RF/microwave filters have become core components to address multi-band compatibility and device miniaturization. They are now a key determinant of front-end modules and application systems. However, in industrial practice, traditional filters still face pain points such as narrow tuning range, high insertion loss, complex control, disconnect between design and verification, and difficulty for novices to get started quickly. This book aims to sort out the types of multimode and tunable filters, combine detailed introductions of each filter from design to verification, establish a complete knowledge system for readers from basic cognition to practical verification, and help solve practical problems in the design of multimode and tunable filters.

The book focuses on the design and verification of multimode and tunable filters in the RF and microwave frequency bands and is divided into seven chapters. Chapter 1 mainly sorts out and introduces the research background and various types of multimode and tunable filters. Chapter 2 focuses on the classification, theoretical derivation, and design verification of filters based on multimode resonators. Chapter 3 classifies and introduces tunable filters with controllable bandwidth. Chapter 4 presents the classification of tunable multi-bandpass filters. Chapter 5 focuses on the classification of tunable bandstop filters. Chapter 6 introduces the tunable frequency selective surface, a device with spatial filtering characteristics. Chapter 7 provides a systematic summary of the entire book and prospects for the future.

Compared with similar books on the market, this book features high systematicness and practicality. It covers the entire process from research background to core theoretical derivations of filter design (such as resonators, coupling coefficients, and external quality factors), combined with specific verification examples. Additionally, the full design process of all filters can be reproduced, allowing readers to conduct independent practice directly based on the design theories and example introductions.

This book is suitable for senior undergraduates, postgraduates majoring in electronic information engineering, communication engineering, microwave technology and other related disciplines, engineers engaged in RF/microwave device design and communication system R&D, as well as researchers focusing on tunable filtering technology. It can also serve as a training textbook for relevant university majors or internal technical training materials for enterprises.

We would like to thank all collaborators who have contributed to the content of this book. Finally, we would like to express our gratitude to every reader who is about to open this book. We hope this book will become your valuable companion in the study and work of tunable multimode RF and microwave filters.

Qingdao, China — Fushun Nian
Qingdao, China — Dinghong Jia
Jiaxing, China — Xiaoguo Huang
Chengdu, China — Shengjun Wei

Contents

Chapter 1
Introduction

1.1 Research Background

Filters are key components in radio frequency (RF) and microwave systems that can selectively process signals of specific frequencies. By designing specific circuits or structures, they allow signals within the target frequency range to pass through smoothly while significantly attenuating or suppressing other unwanted frequency components. Their core function is to precisely separate useful signals from interference signals. For example, in communication systems, they filter out spurious interference from adjacent channels; in radar systems, they suppress noise to improve target detection sensitivity; and in test and measurement, they ensure signal purity to guarantee data accuracy. The importance of filters lies in their role to maintain the performance of RF and microwave systems. Without high-performance filters, systems may suffer from a decrease in signal-to-noise ratio, an increase in communication bit error rate, functional failure, or even complete paralysis due to various interferences. Filters directly determine the signal quality, operational stability, and application reliability of the entire system, and are an indispensable core component in everything from basic electronic devices to complex aerospace systems.

With the evolution of technologies such as 5G/6G communication, the Internet of Things, and cognitive radio, systems with a single frequency band or fixed mode can no longer meet the requirements of multi-standard collaboration, dynamic spectrum access, anti-jamming adaptability, etc. For example, smartphones need to support multiple communication protocols such as LTE, 5G, and Wi-Fi simultaneously, military equipment needs to quickly switch frequency bands to avoid interference, and satellite communication systems need to adapt to the signal characteristics in different orbital environments. This multi-mode and reconfigurable nature require the system to be able to switch seamlessly between different operating modes, frequency bands, or protocols, thereby maximizing the hardware reuse efficiency and reducing costs. This trend has put forward revolutionary requirements for filters, demanding that filters have ultra-wideband or multi-band coverage capabilities, dynamic parameter

F. Nian et al., *RF & Microwave Multi-mode and Tunabe Filters*,
https://doi.org/10.1007/978-981-92-0577-6_1

adjustment capabilities, etc., to cooperate with the entire front-end system to achieve multi-mode reconfigurable operation.

For a reconfigurable front-end system, components such as antennas and low noise amplifiers can adopt broadband and full-frequency coverage characteristics. By using a reconfigurable frequency synthesizer, reconfigurable filters, variable—gain amplifiers, reconfigurable attenuators, etc., a high-performance reconfigurable radio-frequency (RF) system can be achieved. At present, technologies for broadband antennas, broadband low noise amplifiers, variable-gain amplifiers, and tunable attenuators are relatively mature in the market. In broadband transceivers as depicted in Fig. 1.1a, sub-band filters and switches are commonly used to realize the reconfigurable RF filter [1]. This not only increases the volume and development cycle but also reduces the receiver sensitivity. The tunable/reconfigurable RF filter can reduce the complexity and cost of the RF front-end, as shown in Fig. 1.1b. The reconfigurable RF filter has become the main limiting factor for the reconfigurable RF front-end and urgently needs to be broken through. Therefore, this book will focus on discussing the research status of multi-mode/reconfigurable filters.

1.2 Research Status of Multimode/Tunable Filters

Multimode filters are designed by utilizing resonators that support multimode oscillations. Complex filtering functions, such as broadband filtering, multiband filtering, and other advanced functionalities, can be achieved through the control of modes in multimode resonators. Since a single resonator is capable of realizing multiple modes, the number of required resonators is significantly reduced, leading to a substantial reduction in the volume of the filter.

Commonly used multimode resonators for filter design include cavity resonators [2, 3], dielectric resonators [4], substrate integrated waveguide (SIW) resonators [5, 6], and microstrip resonators [7–9]. A typical example of an SIW multimode filter, as illustrated in Fig. 1.2, achieves broadband filtering and multi-channel filtering through the precise control of modes in the SIW multimode resonator [5]. Similarly, a typical microstrip multimode filter, as shown in Fig. 1.3, also realizes broadband and multi-channel filtering via the control of modes in microstrip multimode resonators [8]. Furthermore, switchable or tunable filtering functions can be realized through the flexible control of each mode, enabling adaptive performance adjustment in dynamic communication systems.

Both tunable filters and switchable filters have the characteristic of reconfigurable filtering performance. Here, we collectively refer to them as reconfigurable filtering. Reconfigurable filters can be divided into component-level and circuit-level in terms of reconfigurability. Component-level reconfigurability means achieving applications of different frequencies on a set of filters through components such as tunable capacitors, MEMS switches, and PINs. Circuit-level reconfigurability is to meet different applications by switching directly between different sets of filters through switches like MEMS/PINs. Obviously, the first method can reduce costs and volume,

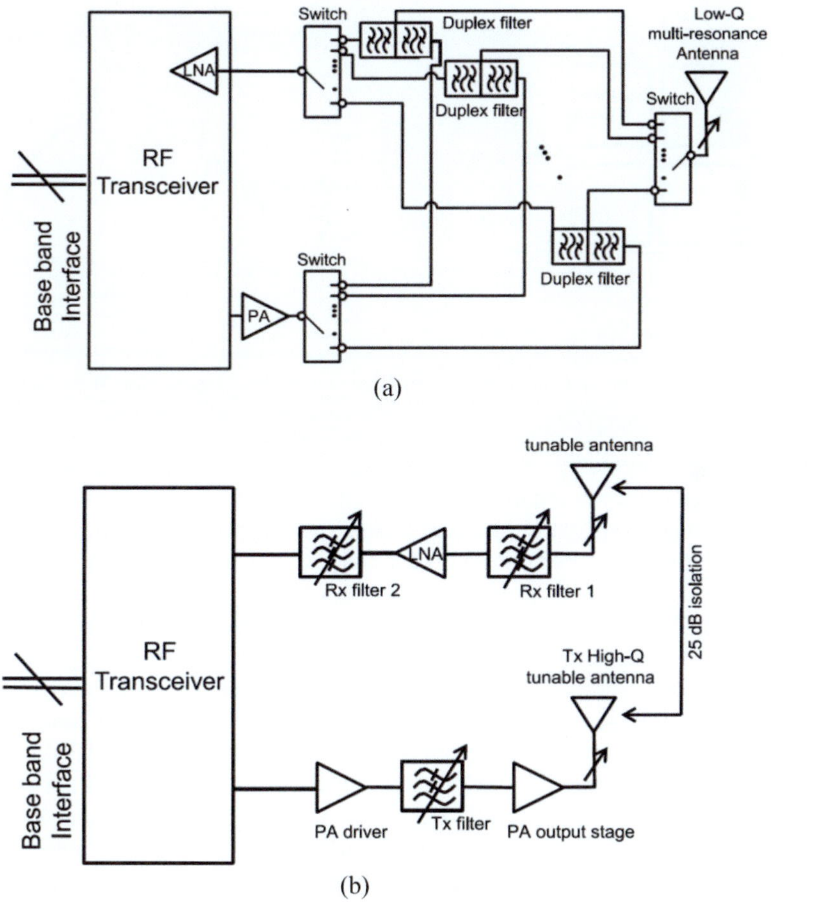

Fig. 1.1 Multi-mode RF front-end. **a** Traditional switchable RF front-end. **b** Proposed tunable RF front-end

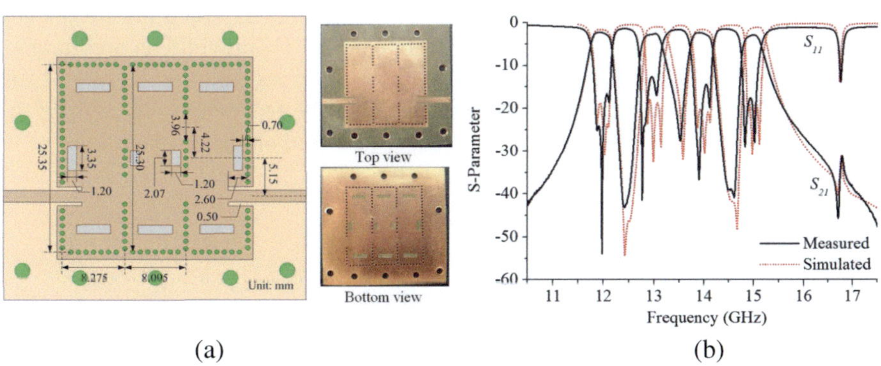

Fig. 1.2 Filter based on multi-mode SIW resonator. **a** Structure diagram. **b** S_{21}/S_{11}

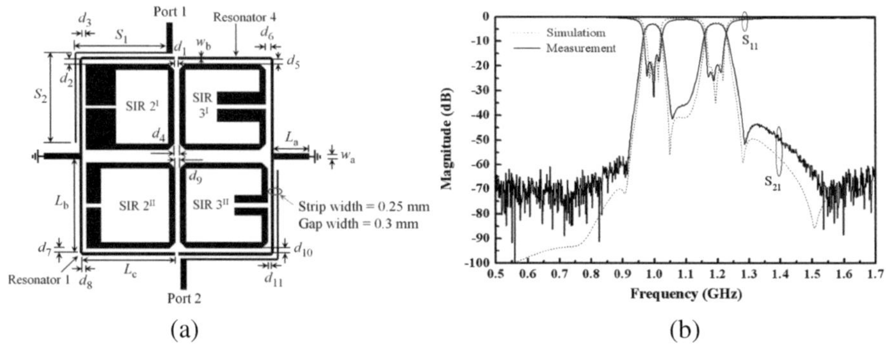

Fig. 1.3 Filter based on multi-mode microstrip resonator. **a** Structure diagram. **b** S_{21}/S_{11}

but the design complexity is extremely high, and currently, various technologies are difficult to meet the requirements of high rectangularity coefficients and wide range adjustment, especially in environments with intensive spectrum utilization. The second method can be tailored to each frequency band, but it has a large volume and high cost.

According to the form of reconfigurability, it can be divided into: (1) Reconfigurability by mechanical deformation of the resonator, mainly achieved by changing the resonant frequency through the deformation of the resonator structure. For example, adding tuning screws, motors, etc. to form reconfigurability; (2) Reconfigurability realized by loading tunable elements on the resonator, including MEMS tunable capacitors [10], switches, semiconductor tunable varactor diodes, PIN switches, liquid crystal tunable capacitors [11], tunable inductors [12], ferroelectric tunable capacitors [13], etc.; (3) Reconfigurability achieved by tuning the substrate and dielectric materials, including ferrite substrates [14], piezoelectric materials [15], yttrium iron garnet (YIG) [16], etc.

According to the type of filtered signal, it can be further divided into reconfigurable lowpass [17], reconfigurable bandpass, reconfigurable bandstop [18], and reconfigurable highpass filters [19]. Reconfigurable filters can be classified into frequency reconfigurable, bandwidth reconfigurable, frequency and bandwidth reconfigurable, dual-frequency reconfigurable, and bandpass/bandstop reconfigurable types from the perspective of reconfigurable research content. The following will analyze the research status of reconfigurable filters one by one based on the research content.

1.2.1 Frequency Reconfigurable Filter

At present, most of the existing literatures are about frequency reconfigurable filters. The focus of the research on frequency reconfigurable filters mainly concerns the pre-settable bandwidth and wide-range reconfigurable characteristics during the frequency reconfigurable process. Park first successfully constructed three second-order tunable filters with pre-settable bandwidth (that is, the relative bandwidth can increase, decrease, or remain unchanged with the change of the center frequency) by using independent electric and magnetic coupling structures [20]. The insertion loss within the tunable range of 850–1300 MHz is 1–3 dB, as shown in Fig. 1.4. Moreover, detailed derivation and design methods were given, which provided theoretical guidance for the design of filters with pre-settable bandwidth and frequency reconfigurability. Most of the subsequent frequency reconfigurable filters with pre-settable bandwidth are based on the electromagnetic hybrid coupling structure.

By taking advantage of the characteristic that the odd and even modes of a multimode resonator are easy to regulate, a bandwidth pre-settable and reconfigurable filter can also be realized. For example, Tang successfully designed two tunable filters with constant absolute bandwidth using a dual-mode resonator loaded with varactor diodes [21]. These filters can generate a transmission zero at the high frequency or low frequency side of the passband respectively, increasing the frequency selection characteristic, as shown in Fig. 1.5. Luo further realized a high performance constant bandwidth tunable filter by using multiple groups of dual-mode resonators loaded with variable capacitors [22].

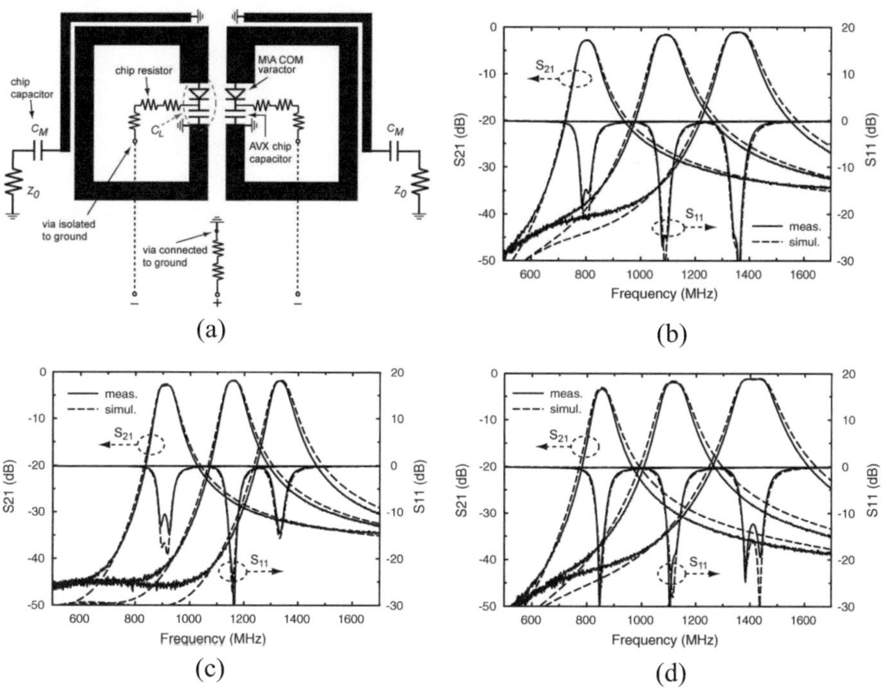

Fig. 1.4 Reconfigurable filter based on independent electrical and magnetic coupling structures. **a** Schematic diagram. **b** Constant relative bandwidth. **c** Constant absolute bandwidth. **d** Increased relative bandwidth

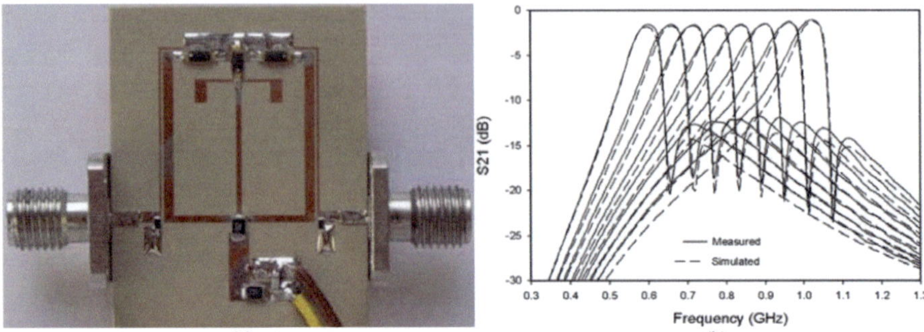

(a) (b)

Fig. 1.5 Tunable filter with constant absolute bandwidth based on dual-mode resonator. **a** Structure diagram. **b** Tunable S_{21}

The evanescent-mode resonator has a very high Q value characteristic. Anand studied the varactor-loaded evanescent-mode resonator and successfully constructed a constant relative-bandwidth tunable filter with a broadband reconfigurable range [23], which has the advantages of a wide adjustment range and low insertion loss. Figure 1.6 indicates the configuration and tunable characteristics of the filter.

Capacitors and inductors have good Q value characteristics and frequency stability at low frequencies. Many LC reconfigurable filters have been widely studied below 1 GHz [24]. Figure 1.7 presents the configuration and tunable characteristics of the LC filter. Lower frequency tunable filter can be realized by helical resonators and variable capacitive loading. The helical filter is manufactured with a CNC process and the tuning is achieved through piezoelectrically actuated tuners, which are bonded over the helix apex [25]. The photograph and tuning characteristics of the tunable helical filter are presented in Fig. 1.8.

The microfluidic reconfigurable filter realizes frequency reconfiguration by having the metallic fluid flow through the microfluidic channels under the resonator, thus equivalently changing the electrical length of the resonator [26]. The photograph and tuning characteristics of the microfluidic reconfigurable filter are presented in Fig. 1.9. This technology has advantages such as simple processing, no frequency

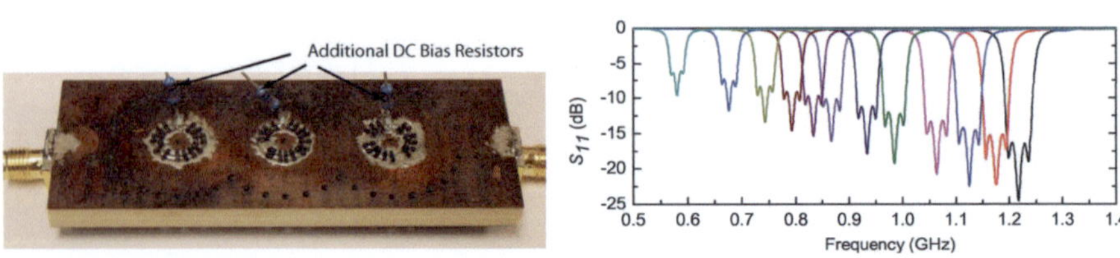

(a) (b)

Fig. 1.6 Tunable filter with constant bandwidth based on evanescent-mode resonator. **a** Structure diagram. **b** Tunable S_{11}

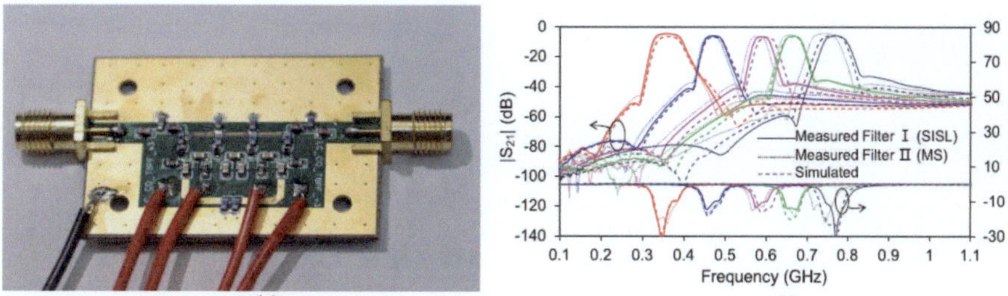

(a) (b)

Fig. 1.7 Frequency reconfigurable filter based on LC structure. **a** Structure diagram. **b** Tunable S_{21}/S_{11}

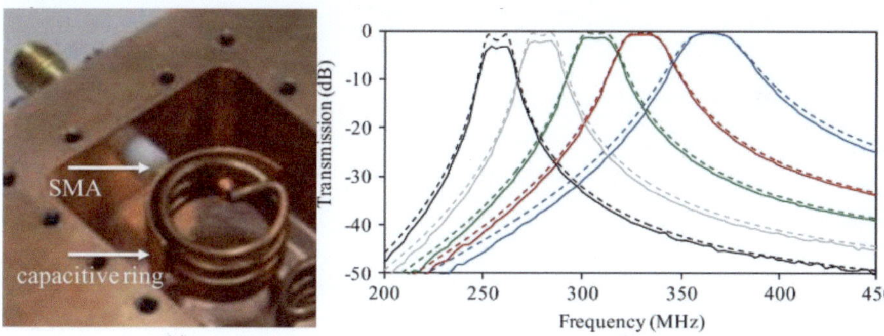

(a) (b)

Fig. 1.8 Frequency reconfigurable filter based on helical resonators. **a** Structure diagram. **b** Tunable S_{21}

limitation, and large power capacity. However, it also has disadvantages such as slow tuning speed and narrow tuning range.

The reconfigurable filter has characteristics such as simple structure and convenient adjustment. However, the topological structure is only for second order or third order characteristics, and it is difficult to expand it to the synthesis of high order filters with a high rectangular coefficient. Therefore, many researchers have proposed filters with high order characteristics [27–29], but they cannot meet the wide range reconfigurable characteristics, as shown in Fig. 1.10 [27].

1.2.2 Bandwidth Reconfigurable Filter

Compared with frequency reconfigurable design, bandwidth reconfigurable design is more complex. For example, to design an N-order frequency reconfigurable bandpass filter, only one type of varactor (N pieces) is needed to adjust the resonant frequency. However, to design an N-order bandwidth reconfigurable bandpass filter, at least (N/

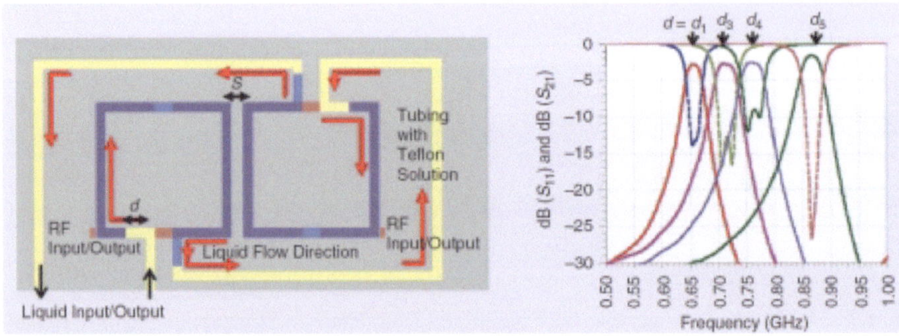

(a) (b)

Fig. 1.9 Frequency reconfigurable filter based on microfluidic technology. **a** Structure diagram. **b** Tunable S_{11}/S_{21}

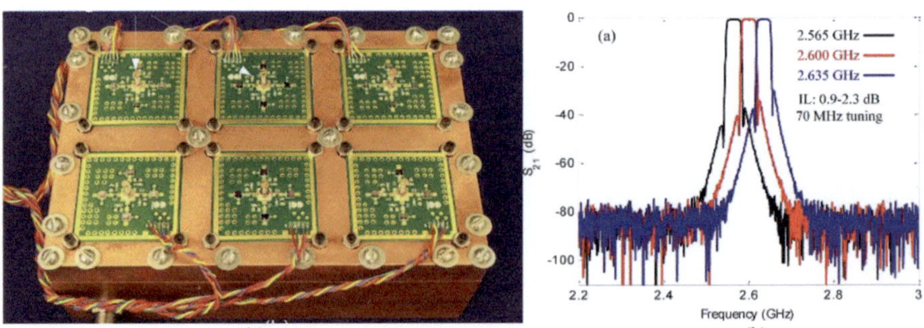

(a) (b)

Fig. 1.10 Sixth-order reconfigurable filter based on MEMS capacitors and piezoelectric materials. **a** Structure diagram. **b** Tunable S_{21}

2 + 1) different capacitances are required to adjust the coupling coefficients. Moreover, some coupling coefficients (magnetic coupling coefficient and weak coupling coefficient) cannot be adjusted by varactors. Varactors can only adjust the electric coupling coefficient. Therefore, the bandpass filters that can use varactors to adjust the coupling coefficients and thus achieve bandwidth reconfiguration are all broadband or ultra broadband filters [30, 31]. A bandwidth reconfigurable filter based on multi-mode resonator is indicated in Fig. 1.11, in which the reconfigurable characteristic is achieved by the PIN switch [30].

In addition, two or more sets of filters can also be used to achieve bandwidth reconfigurability, as depicted in Fig. 1.12a. This method has advantages such as simple design and high rectangularity, but it has the disadvantages of large volume and high cost. Reconfigurable lowpass and highpass filters in series can also be used to achieve bandwidth reconfigurability, as depicted in Fig. 1.12b. This scheme is simple, low in cost, and both the bandwidth and frequency are reconfigurable. However, the Q values of the reconfigurable high-pass or low-pass filters themselves are relatively low, and

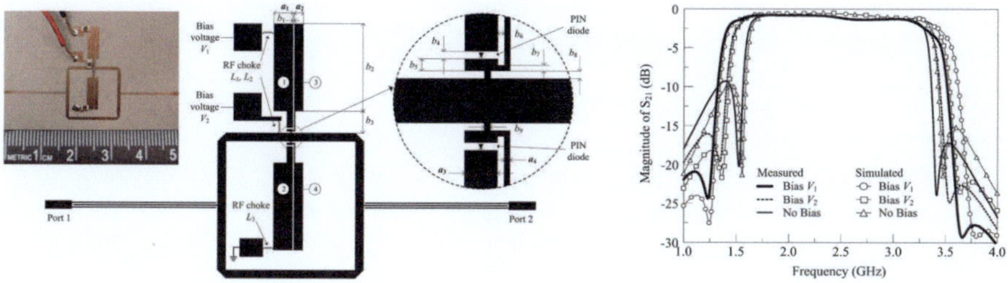

(a) (b)

Fig. 1.11 Bandwidth reconfigurable filter based on multi-mode. **a** Structure diagram. **b** Configurable bandwidth

the roll-off characteristics are not high. Furthermore, two cascade reconfigurable bandpass filter can also be used to achieve bandwidth reconfigurability, as depicted in Fig. 1.12c. This solution also has advantages in volume, cost, and the reconfigurability of both bandwidth and frequency. Moreover, it has good frequency selectivity characteristics. The possible problem is the matching issue during cascading.

1.2.3 Bandwidth and Frequency Reconfigurable Filter

Frequency and bandwidth reconfigurable filters can achieve suitable performance for various communication services. Therefore, this kind of reconfigurable filter is the ultimate design goal of reconfigurable filters. Currently, there are relatively few research results that can simultaneously achieve reconfigurability of both the center frequency and bandwidth. Sanchez-Renedo introduced adjustable coupling attenuators and tunable capacitors in the comb-shaped resonator structure to achieve full tunability of the bandwidth and center frequency respectively, and it has good out of band suppression ability [32]. The photograph and tuning characteristics of the bandwidth and frequency reconfigurable filter are presented in Fig. 1.13. Serrano used a patch type dual-mode resonator loaded with variable capacitors to achieve a second order fully adjustable filter [33]. However, a fully adjustable filter not only needs to adjust the resonator itself but also needs to adjust the coupling coefficients everywhere. Therefore, multiple groups of capacitors need to be used for control, resulting in a relatively complex control strategy and relatively large insertion loss.

1.2.4 Dual-Band Reconfigurable Filter

Dual-band reconfigurable filters can not only expand the range of single-frequency reconfiguration but also realize dual-band simultaneous operation, so research on

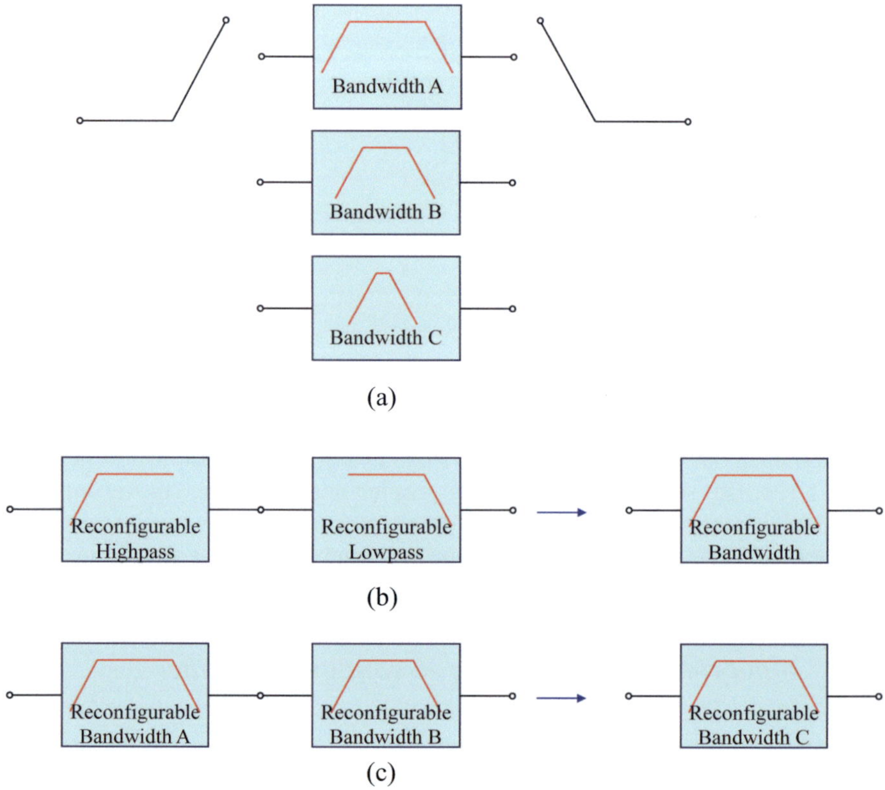

Fig. 1.12 Bandwidth reconfigurable filter based on filter groups. **a** Two or more sets of filters. **b** Reconfigurable highpass and lowpass filters in series. **c** Two or more cascade reconfigurable filters

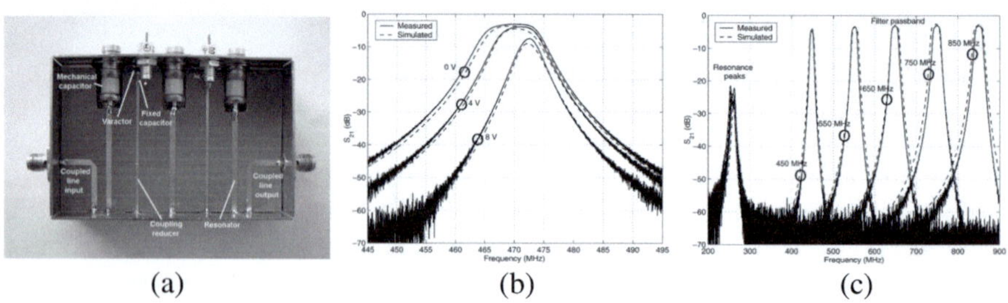

Fig. 1.13 Bandwidth and frequency reconfigurable filter. **a** Structure diagram. **b** Configurable bandwidth. **c** Configurable frequency

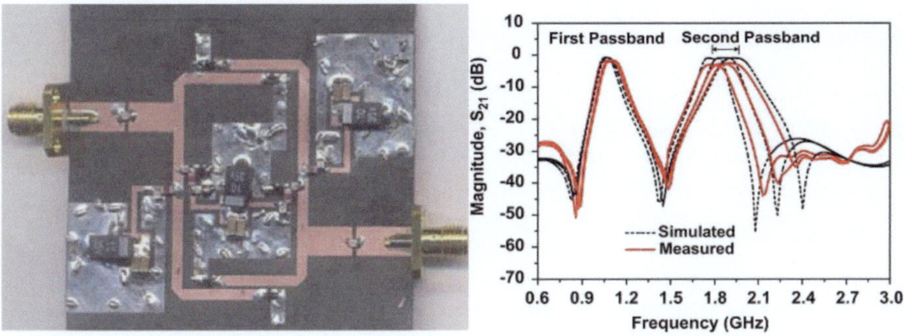

(a) (b)

Fig. 1.14 Dual-band independently reconfigurable filter. **a** Structure diagram. **b** Tunable passband

dual-band reconfigurable filters is of great significance. Djoumessi realized a dual-band independently reconfigurable filter by combining 1/4 wavelength resonators with band-stop filters [34], but this method brings high adjustment complexity and loss due to the need for a large number of varactors; Huang and Chaudhary realized dual-band independently reconfigurable filters using direct coupling of dual-mode resonators and shared input/output structures respectively [35, 36], and the filter with a shared input/output structure realizes dual-band independent reconfiguration by using two independent paths [36], as shown in Fig. 1.14. Similarly, dual-frequency reconfigurable filters also face the problem of synthesizing high-order reconfigurable filters with high rectangular coefficients.

1.2.5 Bandpass-Bandstop Reconfigurable Filter

For environments where the dynamic power changes of signals and interference are relatively large, a bandpass/bandstop switchable filter is required to select the desired signal power level at any time. Chen designed a bandpass/bandstop switchable frequency response using a closed-loop resonator loaded with RF switches [37]. Naglich realized second order and fourth order bandpass/band top switchable filters respectively [38]. Cho achieved the transition between bandpass and bandstop by adjusting the on/off state between the source and the load [39]. The structure and its reconfigurable performance are shown in Fig. 1.15.

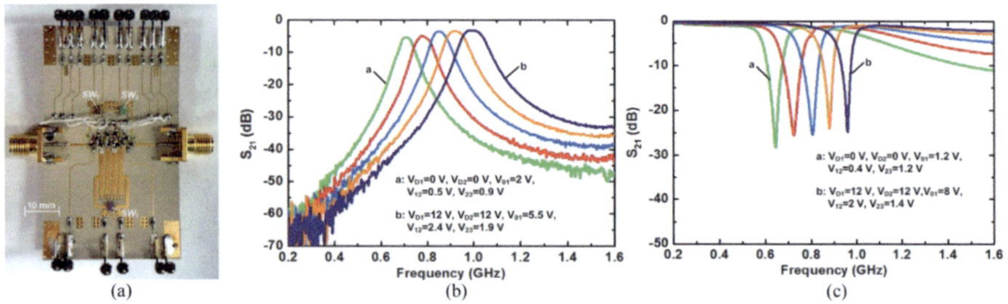

Fig. 1.15 Bandpass-bandstop reconfigurable filter. **a** Structure diagram. **b** Configurable bandpass response. **c** Configurable bandstop response

1.2.6 Other Reconfigurable Filtering Advices

In addition to tunable/reconfigurable filters, there are other devices with adjustable filtering functions, such as tunable frequency selective surfaces [40] and tunable multiplexers [41, 42]. By combining with the design of tunable components, these devices also achieve functions like spatially adjustable filtering and multi-channel controllable filtering. A typical example of a tunable frequency selective surface, as illustrated by Fig. 1.16, realizes a tunable passband through adjustable capacitors [40].

There are also some reports on the integrated design of tunable filters with other functional devices [43–45]. For instance, the integration of tunable filters and amplifiers has been employed to develop tunable filtering amplification front-ends [43]. As shown in Fig. 1.17, a broadband tunable amplification function is realized through the design of a third-order tunable filter combined with an amplifier. Additionally, there are reports on the integration of tunable filters and antennas. As illustrated in Fig. 1.18, the combination of a tunable filter and an antenna enables reconfigurable antenna transceiving over a wide bandwidth by leveraging the tunable filtering function in conjunction with the transceiving function of the antenna [44]. The above-mentioned

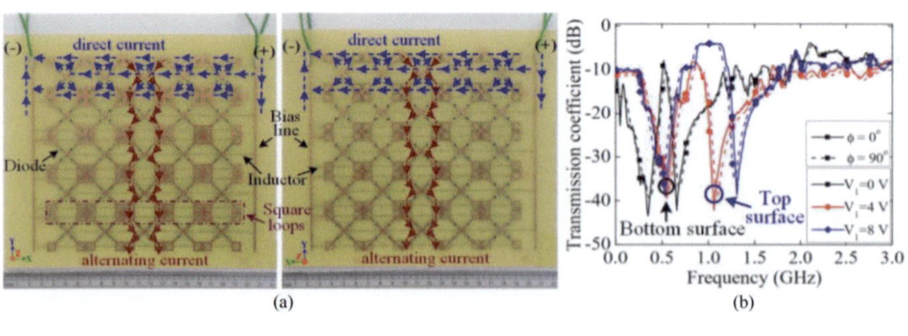

Fig. 1.16 Tunable frequency selective surface. **a** Structure diagram. **b** Tunable transmission coefficient

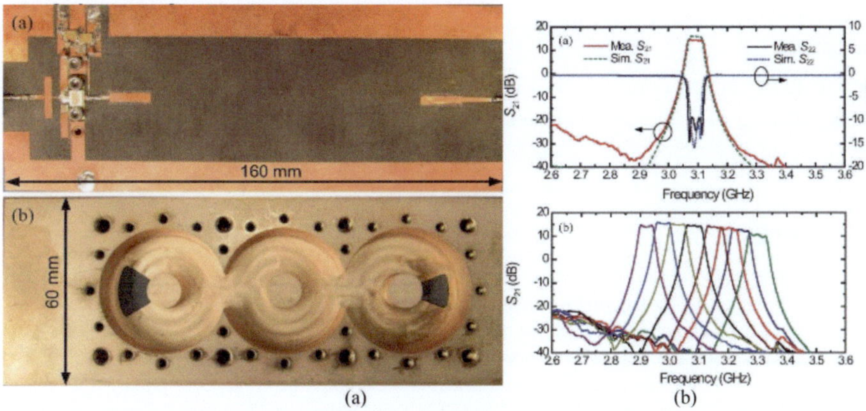

Fig. 1.17 Tunable filter combined with an amplifier. **a** Structure diagram. **b** Configurable S_{21}

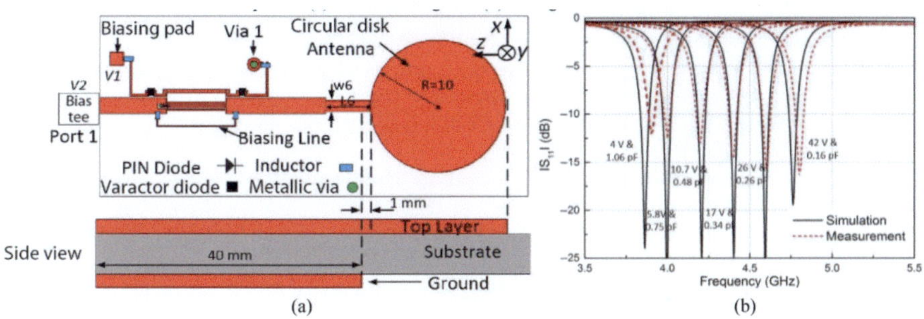

Fig. 1.18 Tunable filter combined with an antenna. **a** Structure diagram. **b** Configurable S_{11}

integrated designs of tunable filters with other front-end systems, on the basis of achieving front-end tunability and reconfigurability, can optimize and improve the overall front-end performance and integration level through joint design.

1.3 Conclusion

With the development of multi-mode/reconfigurable front-end systems, new requirements such as ultra-wideband, multi-mode, and dynamic parameter adjustment have been put forward for filters. The emergence of multi-mode/reconfigurable filters can meet the needs of the above-mentioned systems. By means of controllable structures and tunable components, the functions of multi-mode or reconfigurability can be achieved. This section focuses on summarizing and analyzing the currently mainstream implementation methods of reconfigurable filters, and illustrates the implementation methods of reconfigurable filters in some typical cases, laying a foundation for the writing of subsequent chapters.

References

1. P. Bahramzy et al., "A Tunable RF Front-End With Narrowband Antennas for Mobile Devices," in IEEE Transactions on Microwave Theory and Techniques, vol. 63, no. 10, pp. 3300–3310, Oct. 2015, https://doi.org/10.1109/TMTT.2015.2470237.
2. S. Bastioli, C. Tomassoni and R. Sorrentino, "A New Class of Waveguide Dual-Mode Filters Using TM and Nonresonating Modes," in IEEE Transactions on Microwave Theory and Techniques, vol. 58, no. 12, pp. 3909–3917, Dec. 2010, https://doi.org/10.1109/TMTT.2010.2086068.
3. S. Bastioli and R. V. Snyder, "New Triple-Resonance Configuration Using Stubbed Waveguide Dual-Mode Cavities," in IEEE Microwave and Wireless Components Letters, vol. 32, no. 6, pp. 652–655, June 2022, https://doi.org/10.1109/LMWC.2022.3160299.
4. S. Bastioli and R. V. Snyder, "Inline Pseudoelliptic TE01-Mode Dielectric Resonator Filters Using Multiple Evanescent Modes to Selectively Bypass Orthogonal Resonators," in IEEE Transactions on Microwave Theory and Techniques, vol. 60, no. 12, pp. 3988–4001, Dec. 2012, https://doi.org/10.1109/TMTT.2012.2222659.
5. X. Guo, L. Zhu and W. Wu, "Design Method for Multiband Filters With Compact Configuration in Substrate Integrated Waveguide," in IEEE Transactions on Microwave Theory and Techniques, vol. 66, no. 6, pp. 3011–3018, June 2018, https://doi.org/10.1109/TMTT.2018.2830337.
6. Z. Liu, G. Xiao and L. Zhu, "Triple-Mode Bandpass Filters on CSRR-Loaded Substrate Integrated Waveguide Cavities," in IEEE Transactions on Components, Packaging and Manufacturing Technology, vol. 6, no. 7, pp. 1099–1105, July 2016, https://doi.org/10.1109/TCPMT.2016.2574562.
7. Q. Li, Y. Zhang and C.-T. M. Wu, "High-Selectivity and Miniaturized Filtering Wilkinson Power Dividers Integrated With Multimode Resonators," in IEEE Transactions on Components, Packaging and Manufacturing Technology, vol. 7, no. 12, pp. 1990–1997, Dec. 2017, https://doi.org/10.1109/TCPMT.2017.2706958.
8. C.-F. Chen, T.-M. Shen, T.-Y. Huang and R.-B. Wu, "Design of Multimode Net-Type Resonators and Their Applications to Filters and Multiplexers," in IEEE Transactions on Microwave Theory and Techniques, vol. 59, no. 4, pp. 848–856, April 2011, https://doi.org/10.1109/TMTT.2011.2109392.
9. M. Pal and R. Ghatak, "A Distinctive Resonance: Multiband Bandpass Filter Design Techniques Using Multimode Resonators," in IEEE Microwave Magazine, vol. 16, no. 11, pp. 36–55, Dec. 2015, https://doi.org/10.1109/MMM.2015.2478095.
10. A. Abbaspour-Tamijani, L. Dussopt and G. M. Rebeiz, "Miniature and tunable filters using MEMS capacitors," in IEEE Transactions on Microwave Theory and Techniques, vol. 51, no. 7, pp. 1878–1885, July 2003, https://doi.org/10.1109/TMTT.2003.814317.
11. J. A. Yeh, C. A. Chang, Chih-Cheng Cheng, Jing-Yi Huang and S. S. H. Hsu, "Microwave characteristics of liquid-crystal tunable capacitors," in IEEE Electron Device Letters, vol. 26, no. 7, pp. 451–453, July 2005, https://doi.org/10.1109/LED.2005.851118.
12. Fang, DM., Yuan, Q., Li, XH. et al. Electrostatically driven tunable radio frequency inductor. Microsyst Technol 16, 2119–2122 (2010). https://doi.org/10.1007/s00542-010-1138-1.

13. J. Nath et al., "An electronically tunable microstrip bandpass filter using thin-film Barium-Strontium-Titanate (BST) varactors," in IEEE Transactions on Microwave Theory and Techniques, vol. 53, no. 9, pp. 2707–2712, Sept. 2005, https://doi.org/10.1109/TMTT.2005.854196.
14. G.-M. Yang, O. Obi, G. Wen and N. X. Sun, "Design of Tunable Bandpass Filters With Ferrite Sandwich Materials by Using a Piezoelectric Transducer," in IEEE Transactions on Magnetics, vol. 47, no. 10, pp. 3732–3735, Oct. 2011, https://doi.org/10.1109/TMAG.2011.2155037.
15. S. Saeedi, S. Atash-Bahar, J. Lee and H. Sigmarsson, "Control System for Piezoelectric-Actuator-Based Tunable Evanescent-Mode Cavity Microwave Filters," in IEEE Transactions on Components, Packaging and Manufacturing Technology, vol. 8, no. 11, pp. 1979–1989, Nov. 2018, https://doi.org/10.1109/TCPMT.2018.2874939.
16. G.-M. Yang, J. Wu, J. Lou, M. Liu and N. X. Sun, "Low-Loss Magnetically Tunable Bandpass Filters With YIG Films," in IEEE Transactions on Magnetics, vol. 49, no. 9, pp. 5063–5068, Sept. 2013, https://doi.org/10.1109/TMAG.2013.2253114.
17. J. Ni and J. Hong, "Compact Continuously Tunable Microstrip Low-Pass Filter," in IEEE Transactions on Microwave Theory and Techniques, vol. 61, no. 5, pp. 1793–1800, May 2013, https://doi.org/10.1109/TMTT.2013.2252919.
18. A. Ebrahimi, T. Baum, J. Scott and K. Ghorbani, "Continuously Tunable Dual-Mode Bandstop Filter," in IEEE Microwave and Wireless Components Letters, vol. 28, no. 5, pp. 419–421, May 2018, https://doi.org/10.1109/LMWC.2018.2821841.
19. J. Ni and J. Hong, "Compact Varactor-Tuned Microstrip High-Pass Filter With a Quasi-Elliptic Function Response," in IEEE Transactions on Microwave Theory and Techniques, vol. 61, no. 11, pp. 3853–3859, Nov. 2013, https://doi.org/10.1109/TMTT.2013.2281964.
20. S.-J. Park and G. M. Rebeiz, "Low-Loss Two-Pole Tunable Filters With Three Different Predefined Bandwidth Characteristics," in IEEE Transactions on Microwave Theory and Techniques, vol. 56, no. 5, pp. 1137–1148, May 2008, https://doi.org/10.1109/TMTT.2008.921638.
21. W. Tang and J.-S. Hong, "Varactor-Tuned Dual-Mode Bandpass Filters," in IEEE Transactions on Microwave Theory and Techniques, vol. 58, no. 8, pp. 2213–2219, Aug. 2010, https://doi.org/10.1109/TMTT.2010.2052958.
22. X. Luo, S. Sun and R. B. Staszewski, "Tunable Bandpass Filter With Two Adjustable Transmission Poles and Compensable Coupling," in IEEE Transactions on Microwave Theory and Techniques, vol. 62, no. 9, pp. 2003–2013, Sept. 2014, https://doi.org/10.1109/TMTT.2014.2337287.
23. A. Anand, J. Small, D. Peroulis and X. Liu, "Theory and Design of Octave Tunable Filters With Lumped Tuning Elements," in IEEE Transactions on Microwave Theory and Techniques, vol. 61, no. 12, pp. 4353–4364, Dec. 2013, https://doi.org/10.1109/TMTT.2013.2287674.
24. L. Liu, Q. Xiang and Q. Feng, "A Self-Packaged Tunable LC Fourth-Order Filter With Cross-Coupling and Bandwidth Compensation," in IEEE Microwave and Wireless Technology Letters, vol. 35, no. 4, pp. 420–423, April 2025, https://doi.org/10.1109/LMWT.2025.3535706.
25. D. Psychogiou and D. Peroulis, "Tunable VHF Miniaturized Helical Filters," in IEEE Transactions on Microwave Theory and Techniques, vol. 62, no. 2, pp. 282–289, Feb. 2014, https://doi.org/10.1109/TMTT.2013.2294592.
26. K. Entesari and A. P. Saghati, "Fluidics in Microwave Components," in IEEE Microwave Magazine, vol. 17, no. 6, pp. 50–75, June 2016, https://doi.org/10.1109/MMM.2016.2538513.
27. S. Fouladi, F. Huang, W. D. Yan and R. R. Mansour, "High-Q Narrowband Tunable Combline Bandpass Filters Using MEMS Capacitor Banks and Piezomotors," in IEEE Transactions on Microwave Theory and Techniques, vol. 61, no. 1, pp. 393–402, Jan. 2013, https://doi.org/10.1109/TMTT.2012.2226601.
28. D. Bouyge et al., "Reconfigurable 4 pole bandstop filter based on RF-MEMS-loaded split ring resonators," 2010 IEEE MTT-S International Microwave Symposium, Anaheim, CA, USA, 2010, pp. 588–591, https://doi.org/10.1109/MWSYM.2010.5517736.
29. A. Iqbal, A. W. Ahmad, A. Smida and N. K. Mallat, "Tunable SIW Bandpass Filters With Improved Upper Stopband Performance," in IEEE Transactions on Circuits and Systems II: Express Briefs, vol. 67, no. 7, pp. 1194–1198, July 2020, https://doi.org/10.1109/TCSII.2019.2936495.

30. A.H. Kim and K. Chang, "Ring Resonator Bandpass Filter With Switchable Bandwidth Using Stepped-Impedance Stubs," in IEEE Transactions on Microwave Theory and Techniques, vol. 58, no. 12, pp. 3936–3944, Dec. 2010, https://doi.org/10.1109/TMTT.2010.2084584.
31. X. Huang, Q. Feng and Q. Xiang, "Bandpass Filter With Tunable Bandwidth Using Quadruple-Mode Stub-Loaded Resonator," in IEEE Microwave and Wireless Components Letters, vol. 22, no. 4, pp. 176–178, April 2012, https://doi.org/10.1109/LMWC.2012.2188280.
32. M. Sanchez-Renedo, R. Gomez-Garcia, J. I. Alonso and C. Briso-Rodriguez, "Tunable combline filter with continuous control of center frequency and bandwidth," in IEEE Transactions on Microwave Theory and Techniques, vol. 53, no. 1, pp. 191–199, Jan. 2005, https://doi.org/10.1109/TMTT.2004.839309.
33. Serrano, A.L.C. and Correra, F.S. (2007), Dual-mode bandpass filter using triangular patch resonator with inverted T-shaped slot. Microw. Opt. Technol. Lett., 49: 2897–2902. https://doi.org/10.1002/mop.22927.
34. E. E. Djoumessi, M. Chaker, and K. Wu, Varactor-tuned quarter-wavelength dual-bandpass filter, IET Microwaves, Antennas & Propagation, 2009, 3(1): 117–124.
35. X. Huang, L. Zhu, Q. Feng, Q. Xiang and D. Jia, "Tunable Bandpass Filter With Independently Controllable Dual Passbands," in IEEE Transactions on Microwave Theory and Techniques, vol. 61, no. 9, pp. 3200–3208, Sept. 2013, https://doi.org/10.1109/TMTT.2013.2273894.
36. G. Chaudhary, Y. Jeong and J. Lim, "Harmonic Suppressed Dual-Band Bandpass Filters With Tunable Passbands," in IEEE Transactions on Microwave Theory and Techniques, vol. 60, no. 7, pp. 2115–2123, July 2012, https://doi.org/10.1109/TMTT.2012.2197020.
37. Y.-M. Chen, S.-F. Chang, C.-Y. Chou and K.-H. Liu, "A reconfigurable bandpass-bandstop filter based on varactor-loaded closed-ring resonators [Technical Committee]," in IEEE Microwave Magazine, vol. 10, no. 1, pp. 138–140, February 2009, https://doi.org/10.1109/MMM.2008.930742.
38. E. J. Naglich, J. Lee, D. Peroulis and W. J. Chappell, "A Tunable Bandpass-to-Bandstop Reconfigurable Filter With Independent Bandwidths and Tunable Response Shape," in IEEE Transactions on Microwave Theory and Techniques, vol. 58, no. 12, pp. 3770–3779, Dec. 2010, https://doi.org/10.1109/TMTT.2010.2086533.
39. Y.-H. Cho and G. M. Rebeiz, "0.7–1.0-GHz Reconfigurable Bandpass-to-Bandstop Filter With Selectable 2- and 4-Pole Responses," in IEEE Transactions on Microwave Theory and Techniques, vol. 62, no. 11, pp. 2626–2632, Nov. 2014, https://doi.org/10.1109/TMTT.2014.2359645.
40. S. Ghosh and K. V. Srivastava, "A Dual-Band Tunable Frequency Selective Surface With Independent Wideband Tuning," in IEEE Antennas and Wireless Propagation Letters, vol. 19, no. 10, pp. 1808–1812, Oct. 2020, https://doi.org/10.1109/LAWP.2020.3019584.
41. C. Arnold, J. Parlebas, R. Meiser and T. Zwick, "Fully Reconfigurable Manifold Multiplexer," in IEEE Transactions on Microwave Theory and Techniques, vol. 65, no. 10, pp. 3885–3891, Oct. 2017, https://doi.org/10.1109/TMTT.2017.2693154.
42. A.C. Guyette, "Theory and Design of Intrinsically Switched Multiplexers With Optimum Phase Linearity," in IEEE Transactions on Microwave Theory and Techniques, vol. 61, no. 9, pp. 3254–3264, Sept. 2013, https://doi.org/10.1109/TMTT.2013.2274963.
43. K. Chen, T.-C. Lee and D. Peroulis, "Co-Design of Multi-Band High-Efficiency Power Amplifier and Three-Pole High-Q Tunable Filter," in IEEE Microwave and Wireless Components Letters, vol. 23, no. 12, pp. 647–649, Dec. 2013, https://doi.org/10.1109/LMWC.2013.2283876.
44. P.-Y. Qin, F. Wei and Y. J. Guo, "A Wideband-to-Narrowband Tunable Antenna Using A Reconfigurable Filter," in IEEE Transactions on Antennas and Propagation, vol. 63, no. 5, pp. 2282–2285, May 2015, https://doi.org/10.1109/TAP.2015.2402295.
45. T. Liang, Y. Dong, Z. Wang, X. Chen and H. Wang, "Quasi-Reflectionless Tunable Filtering Antenna for Multicarrier Transceiver," in IEEE Antennas and Wireless Propagation Letters, vol. 20, no. 6, pp. 1053–1057, June 2021, https://doi.org/10.1109/LAWP.2021.3070806.

Chapter 2
High Selectivity Bandpass Based on Multi-mode Resonators

2.1 Introduction

With the rapid development of wireless communication technologies, microwave filters featuring multiband operation, high selectivity, and compact dimensions have become crucial components for multimode communication systems [1–3]. This chapter focuses on two primary approaches to multimode filter design: microstrip structures and substrate-integrated waveguide (SIW), each offering distinct advantages in terms of integration flexibility/miniaturization and harmonic suppression, respectively.

Among implementations based on microstrip structures, three typical methods prevail. The first method combines two sets of resonators operating at different frequency bands through common input/output ports [4–7]. Such resonators can be realized using either multimode resonators [4, 5] or single-mode resonators [6, 7]. Reference [4] employs double trisection-mode resonators for high selectivity, while Refs. [6, 7] achieve high selectivity via cross-coupling techniques in single-mode resonators. These filters offer dual independent passbands, high selectivity, and straightforward design processes, but require two filter circuits, leading to increased size. The second method realizes dual-band characteristics by precisely controlling the second harmonic or first-order harmonic within a single resonator structure [8–14]. References [8–13] implement dual-band filters using dual-mode resonators with controllable first-order harmonics, while Ref. [14] designs a single-mode dual-band filter using an E-shaped stepped-impedance resonator (SIR). Despite their compact size, these designs generally exhibit lower selectivity compared to the first method, and the performance of each passband is often constrained by the first passband. The third method employs resonators formed by open- or short-circuited microstrip structures [15–17]. These resonators create an additional bandpass response within the intermediate passband region of a wideband bandpass filter. While offering excellent roll-off selectivity, this approach typically entails complex design procedures and

F. Nian et al., *RF & Microwave Multi-mode and Tunabe Filters*,
https://doi.org/10.1007/978-981-92-0577-6_2

results in significantly larger circuit areas due to the multitude of cascaded microstrip elements.

SIW technology has garnered significant attention in the development of microwave and millimeter-wave integrated circuits and systems owing to its compatibility with planar circuit integration, low insertion loss, and ease of fabrication [18]. SIW-based multiband filters have been applied in modern communication systems evolving towards multimode operation and high integration density [19–26]. These filters primarily function via coupling mechanisms between different resonators. Coupling between SIW resonators is typically achieved through inductive vias, magnetic irises [27], or by incorporating evanescent-mode waveguide sections within the same resonator layer [28, 29]. During this coupling process, the filter's stopband or out-of-band rejection performance may degrade to some extent. To address this issue, researchers have proposed solutions involving feed or excitation port design techniques to suppress undesirable higher-order modes in SIW resonators [30, 31]. Furthermore, utilizing multilayer SIW topologies [32] provides greater design freedom for tailoring coupling paths and strengths between resonators while maintaining a compact physical footprint. Design strategies for SIW multimode filters can be categorized into three main types: aperture-loading methods [19, 20], coupling matrix synthesis methods [21, 22], and multi-mode dual-band SIW resonator designs [23–26].

This chapter is structured as follows: Sect. 2.2 details various multimode filter topologies based on microstrip and SIW structures, along with their working principles. Section 2.3 presents theoretical analysis, deriving mathematical models for key parameters and control equations to elucidate the conditions for multimode filter realization. Section 2.4 provides comprehensive design examples accompanied by simulation and measurement results to validate the effectiveness of the proposed methods, topologies, and parametric theories. Finally, Sect. 2.5 summarizes the main findings and contributions.

2.2 Proposed Topology of High Selectivity Bandpass Filters Based on Multi-mode Resonators

2.2.1 Planar Microstrip Dual-Mode and Tri-Mode Resonators-Based Filters

Based on the demand for compact, highly selective multi-band filters discussed in the introduction, we present several innovative topological frameworks in this section to achieve highly selective bandpass filters using multimode resonators. The core design principle of these topologies lies in the rational configuration of the resonator's geometric shape and coupling mechanism within an efficient physical layout. This aims to generate the desired resonant modes and strategically place transmission

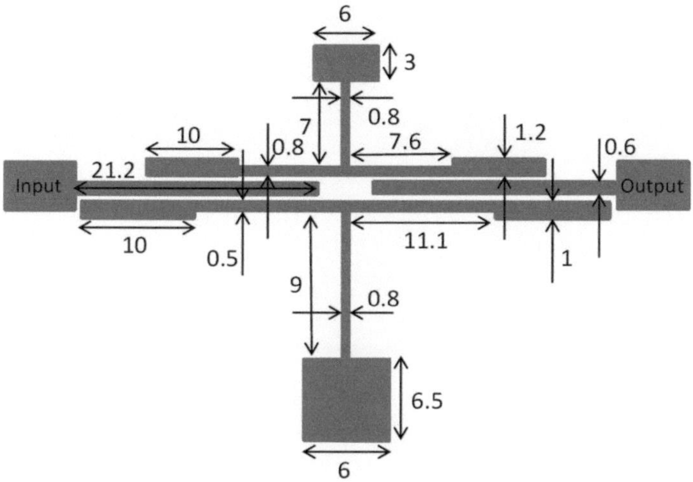

Fig. 2.1 Schematic of the proposed dual-band filter I based on two dual-mode resonators

zeros (TZs), thereby enhancing passband selectivity without introducing excessive circuit complexity.

Initial exploration focused on planar microstrip solutions, where a common I/O feeding scheme was employed to realize a multimode filter with structural compactness. As illustrated in Fig. 2.1, this structure integrates two independently tunable dual-mode resonators formed in a stepped-impedance (SIR) configuration [33]. Each resonator is centered by loading with shunt open-ended stubs serving to shape their resonant characteristics. These two dual-mode resonators share common input/output (I/O) feed lines. This arrangement not only simplifies the structure but also enables independent and flexible design and control of the filter's first and second passbands via their corresponding resonators. By inherently leveraging its coupling mechanism, this structure effectively generates multiple TZs, significantly enhancing the roll-off rate at the edges of both passbands and improving the isolation between them.

To further enhance the selectivity at the high-frequency edge of the first passband, our research perspective broadens to explore the potential of introducing additional modes. The core principle of this design approach lies in the three-dimensional integration of a coplanar waveguide (CPW) ground plane structure with microstrip stepped-impedance resonators (SIRs). This integration enables the creation of an additional transmission zero, specifically targeting the high-end selectivity of the passband without significantly increasing the circuit dimensions [34]. Figure 2.2 presents the proposed tri-mode resonator scheme. The conventional microstrip dual-mode structure is integrated with an underlying CPW structure through a single via. The introduction of the CPW structure contributes an independent third even-mode resonance point, successfully expanding the dual-mode characteristic of the microstrip structure into a tri-mode operation within the same physical footprint.

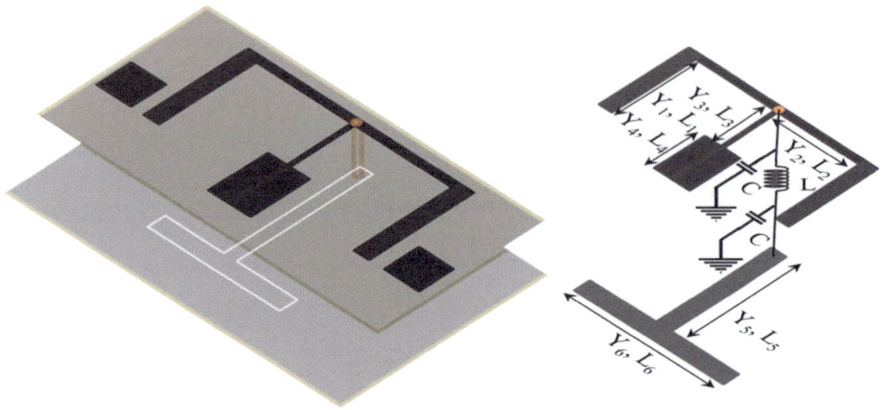

Fig. 2.2 Structure and equivalent circuit mode of the proposed tri-mode resonator

This structural evolution from microstrip to CPW essentially transforms a dual-mode resonator into a compact tri-mode resonator. The interaction between these three modes themselves and their coupling paths with the I/O ports enables the generation of specific TZs near the passband. This significantly improves the roll-off selectivity of the passband, particularly for the first passband.

The microstrip layer adopts an asymmetric stepped-impedance design. Within this structure, the high-impedance section (L_1) and low-impedance section (L_2) collectively form the fundamental resonance mode. The centrally loaded microstrip open stub (L_3) excites a secondary mode primarily through electric-field perturbation. Concurrently, the CPW unit (L_5) etched in the ground layer establishes a third controllable mode, coupled electromagnetically to the upper layers through the vias. Under odd-mode excitation, the symmetrical plane establishes a virtual short-circuit plane. This configuration causes via the inductance and the CPW unit to form a series resonance circuit. Under even-mode excitation, the symmetrical plane instead forms a virtual open-circuit point. In this state, the microstrip stub (L_3) and the CPW unit (L_5) collectively generate a parallel resonance.

Building upon the three-dimensional hybrid structure, this design offers unique frequency tuning advantages. Adjusting the admittance (Y_4) of the microstrip open stub enables independent control of the lower-sideband even-mode frequency, while modifying the length (L_6) of the CPW unit selectively tunes the upper-sideband even-mode frequency. This characteristic arises from the directional perturbation of the even-mode electric field distribution induced by the loaded stubs.

Building upon the preceding advancements, we introduce an additional tri-mode dual-passband filter topology based on the hybrid integration of microstrip lines and CPW [33]. This structure features metallic vias vertically penetrating the dielectric substrate, integrating a microstrip open-ended stub loaded at the center of the microstrip layer with an etched CPW open resonant unit in the ground plane layer. As depicted in Fig. 2.3, this hybrid resonator, interconnected through the metallic vias, evolves the traditional two-dimensional microstrip resonator into a composite cavity supporting three controllable resonant modes.

As illustrated in Fig. 2.4, two identical tri-mode resonators are connected in a back-to-back configuration [34]. The input/output feed lines employ a tapered impedance matching design, establishing a hybrid feeding mechanism combining magnetic and electric coupling within the framework of a common port, thereby realizing a dual-passband filter based on tri-mode resonators.

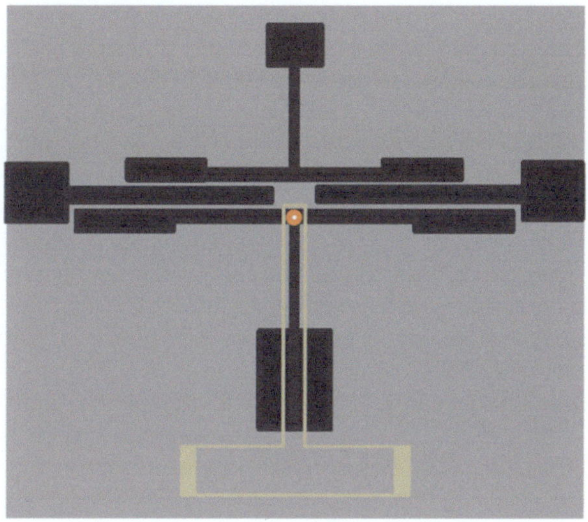

Fig. 2.3 Schematic of the proposed dual-band filter II based on dual-mode resonator and triple-mode resonator

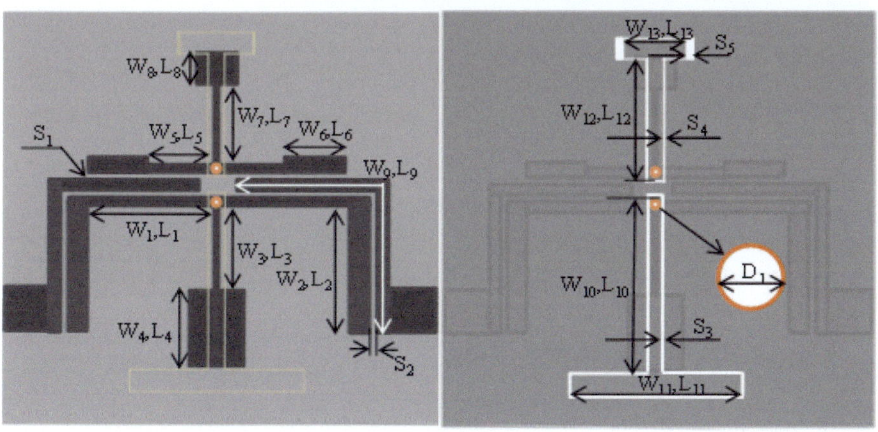

(a) (b)

Fig. 2.4 Schematic of the proposed dual-band filter III based on triple-mode resonators. **a** Top view. **b** Bottom view

2.2.2 *Substrate-Integrated Waveguide: Multimode Resonators-Based Filters*

To extend multi-band filtering capabilities beyond dual-band operations and address the limitations of planar structures in terms of mode density and independent control, we expand this design concept to multilayer SIW technology. SIW offers high-quality-factor resonances and facilitates vertical integration. Herein, a vertical coupling architecture is proposed utilizing apertures etched in the intermediate metal layer between stacked dielectric substrates [35]. This topology, through three-dimensional resonator stacking and hybrid coupling mechanisms, promotes distinct magnetic/electric coupling between adjacent rectangular SIW cavities in neighbouring layers. This successfully resolves the inherent conflict in traditional single-layer filter structures between suppressing higher-order modes and achieving compact design. The key innovation lies in the precise design of these vertical couplings, specifically targeting the fundamental mode (TE_{101}) and the critical first higher-order mode (TE_{102}). By intentionally designing the coupling strength for the TE_{102} mode to be weaker or configuring its coupling type (magnetic/electric) as antagonistic, its excitation can be effectively suppressed. This topology offers a dual advantage: achieving better out-of-band rejection by suppressing spurious passbands caused by TE_{102}, while simultaneously realizing a more compact footprint compared to equivalent single-layer SIW filters employing lateral couplings.

As depicted in Fig. 2.5, two square resonant cavities are aligned vertically. An intermediate layer embedded with reconfigurable metal apertures serves as electromagnetic coupling paths. Precise control over the spatial distribution of rectangular slots and circular apertures in this intermediate layer enables both magnetic coupling (Filter IV) and hybrid electric–magnetic coupling (Filter V). In the magnetic coupling topology (Filter IV), electromagnetic energy efficiently couples the TE_{101} mode via the slot. However, due to its orthogonal magnetic field distribution, the TE_{102} mode experiences minimal coupling strength and cannot be effectively excited. Building on this, the hybrid topology with coupled cancellation effect (Filter V) features a main circular aperture (radius Er_1) etched at the center of the intermediate metal layer—precisely the region of maximum electric field intensity for the fundamental TE_{101} mode, establishing the primary electric coupling path. Concurrently, auxiliary circular apertures (radius Er_2) are symmetrically positioned at the cavity corners, corresponding to the electric field extremal points of the TE_{102} higher-order mode. Adjusting the size of auxiliary apertures does not affect the coupling strength of the fundamental TE_{101} mode, but generates a magnetic coupling component that counteracts the magnetic coupling generated by the main aperture specifically for TE_{102}. Crucially, when Er_2 is tuned to a specific value, these two coupling effects precisely cancel each other at the TE_{102} resonance frequency, neutralizing and dissipating the higher-order mode energy before resonance can occur.

Building upon the proven success of vertical coupling for mode control in SIW, a quad-band filter topology with independently controllable bandwidths is proposed here [36]. This structure employs a single large rectangular SIW cavity, leveraging

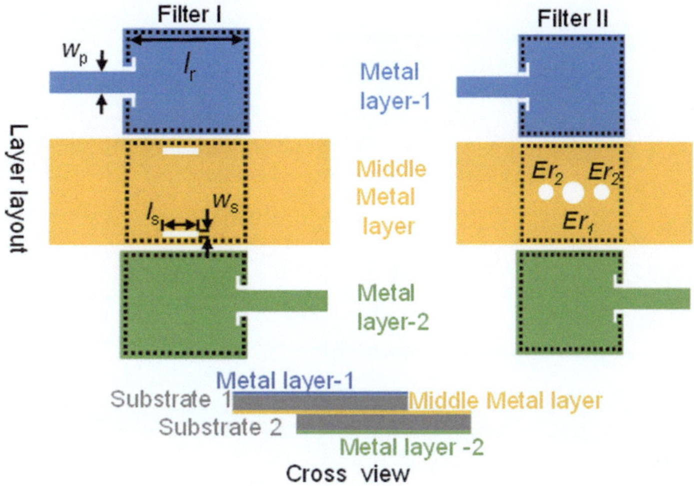

Fig. 2.5 Typical topologies of the proposed vertically coupled multilayer filters: Filter IV (magnetic coupling) and Filter V (electric coupling)

its orthogonally distributed family of TE modes as the foundation for constructing multiple passbands. Specifically, the first four modes (TE_{101}, TE_{102}, TE_{103}, and TE_{104}) are utilized to generate four independent passbands. As shown in Fig. 2.6, the filter utilizes a multilayer vertical stack comprising two dielectric layers sandwiched between three metal layers. Both mode coupling and source-load coupling are implemented on the intermediate metal layers. These intermediate layers are not continuous but are precision-etched with coupling apertures of specific shapes and distributions (including multiple slots s1, s2, s3, s4, and circular holes c1, c2, c3).

Energy coupling for the TE_{101} mode (1st passband) primarily relies on slots s1 and s2 etched in the intermediate layers. The location of these slots is designed based on the electric and magnetic field intensity maxima of TE_{101}, enabling efficient magnetic and electric coupling respectively, and providing the fundamental coupling path. Magnetic coupling for the TE_{102} mode (2nd passband) is mainly achieved through the central circular hole c1, while its electric coupling is facilitated by additional circular holes c2 and c3. This separated coupling design aims to provide relatively independent tuning capability for the second passband. The TE_{103} mode (3rd passband), being a higher-order mode, utilizes circular holes c1, c2, c3 for its electric coupling, while its magnetic coupling relies primarily on c2, c3 and the newly introduced slots s3 and s4. These slots s3 and s4 are specifically designed as coupling interfaces targeting TE_{103} and the next mode. Electric coupling for the TE_{104} mode (4th passband) primarily acts on slot s4. Due to its small size, the electric coupling contribution from s4 is weak. More crucially, its magnetic coupling path is completed via the combined action of c1, c2, c3, and s3.

This vertical coupling architecture enables relatively independent excitation and coupling for each resonant mode. By adjusting the apertures (slots and circular holes)

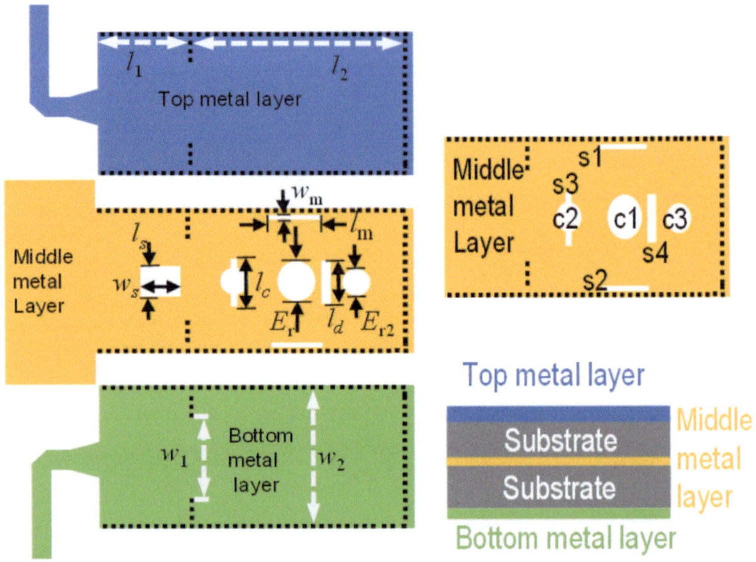

Fig. 2.6 Structure of the proposed quad-band multilayer SIW filter: Filter VI

located at different spatial positions and possessing different shapes, the field characteristics of different modes (specifically, the location and orientation of their field intensity maxima) can be individually targeted for coupling design. This intentional spatial layout and physical isolation allow the bandwidths of the four passbands to be relatively independently adjusted, satisfying the practical system requirements for different bandwidth specifications across different frequency bands.

Direct energy coupling between the source and load plays a critical role in the frequency response. Combined with the multi-mode resonance mechanism described above, it generates multiple TZs near the band-edges (primarily the lower and upper edges) of each passband. These TZs can be positioned very close to the passband edges, resulting in extremely sharp roll-off characteristics that effectively suppress out-of-band interfering signals.

In summary, this section outlines a series of multi-mode resonator-based filter topologies, progressing from planar microstrip structures to SIW architectures. All schemes are unified by a core principle: exploiting and precisely controlling multiple resonant modes. Each proposed structure—from the dual-mode and tri-mode microstrip-CPW resonators, through the multilayer SIW filters, to the final quad-band SIW cavity—represents a distinct architectural approach towards achieving high selectivity and compactness in multi-band filters. The fundamental design strategies revolve around resonator reconfiguration (introducing stubs, CPW elements), novel coupling mechanisms (vertical apertures in multilayer structures), and the targeted utilization of specific resonant modes within complex structures.

2.3 Theoretical Analysis

2.3.1 *Theoretical Analysis of BPF Based on Microstrip Multi-mode Resonators*

Figure 2.7a shows the basic structure of a dual-mode resonator. It can be divided using the symmetrical plane, one of which is regarded as an open circuit or a short circuit for even or odd modes, respectively. In the even mode, $Y_3/2$ and $Y_4/2$ result from dividing the stepped-impedance stubs in half along the plane of symmetry, as shown in Fig. 2.7b. In the odd mode, because the plane of symmetry can be considered as ground plane, a simple equivalent circuit is made as shown in Fig. 2.7c.

The resonance frequencies can be calculated when $\mathrm{Im}[Y_{in\_even}] = 0$ and $\mathrm{Im}[Y_{in\_odd}] = 0$ from the one end of the even- and odd-mode circuit, respectively. The resonance frequency of even mode can be expressed by

$$Y_2\frac{Y_3(Y_4\tan\theta_4 + Y_3\tan\theta_3) + 2Y_2(Y_3 - Y_4\tan\theta_4\tan\theta_3)\tan\theta_2}{2Y_2(Y_3 - Y_4\tan\theta_4\tan\theta_3) - Y_3(Y_4\tan\theta_4 + Y_3\tan\theta_3)\tan\theta_2} + Y_1\tan\theta_1 = 0 \tag{2.1}$$

The resonance frequency of odd mode can be expressed by

$$-jY_2\cot\theta_2 + jY_1\tan\theta_1 = 0 \tag{2.2}$$

The TZ frequencies are obtained when $Y_{21} = Y_{12} = 0$, where the admittance matrices are calculated by Y-parameter between port 1 and port 2 in Fig. 2.7a.

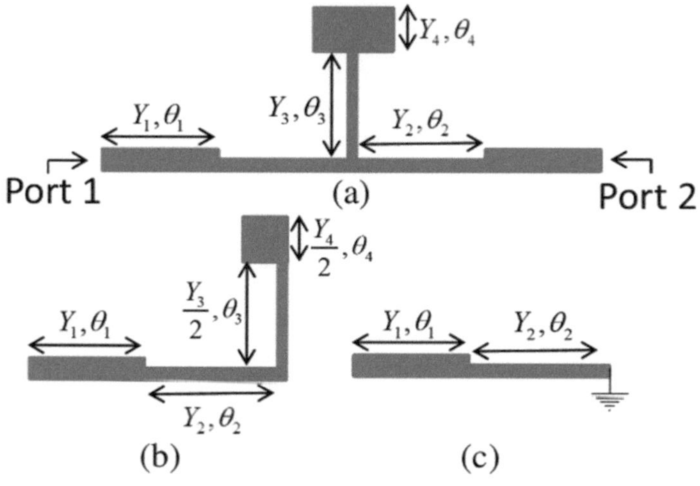

Fig. 2.7 Proposed dual-mode resonator. **a** Basic structure. **b** Even-mode excitation equivalent circuit. **c** Odd-mode excitation equivalent circuit

Assuming $Y_1 = Y_2 = Y_0$ and $\theta_1 + \theta_2 = \theta_0$, the filter TZs can be calculated by

$$Y_{12} = Y_{21} = \frac{-1}{j\frac{1}{Y_1}\cos\theta_0 \sin\theta_0 + j\frac{1}{Y_1}\sin\theta_0\left(\cos\theta_0 + j\frac{Y_{load}}{Y_1}\right)} = 0 \tag{2.3}$$

where $Y_{load} = \frac{jY_3 \sin\theta_3 \cos\theta_4 + jY_4 \cos\theta_3 \sin\theta_4}{\cos\theta_3 \cos\theta_4 - \frac{Y_4}{Y_3}\sin\theta_3 \sin\theta_4}$. Thus, $Y_{load} \rightarrow \infty$ can be deduced and

$$Y_3 - Y_4 \tan\theta_3 \tan\theta_4 = 0 \tag{2.4}$$

From (2.1), (2.2) and (2.4), it can be observed that the odd-mode resonant frequency depends on Y_1, Y_2, θ_1 and θ_2, and the even-mode resonant frequency depends on Y_1, Y_2, Y_3, Y_4, θ_1, θ_2, θ_3 and θ_4. It also can be seen that the TZ just depends on Y_3, Y_4, θ_3 and θ_4. To verify the theoretical analysis, a full-wave electromagnetics simulation was carried out using Sonnet. Figure 2.8 shows the simulated S_{21}-magnitude of a weak coupling resonator circuit according to different Y_{1b}, θ_{4b}, Y_{1t} and θ_{4t}, where the subscript b and t represent the bottom dual-mode resonator and the top dual-mode resonator. As Y_{1b} or Y_{1t} is varied, the odd- and even-mode resonant frequencies are tuned simultaneously, whereas θ_{4b} or θ_{4t} is varied, the odd-mode resonant frequency is fixed and the even-mode resonant frequency can be tuned. Thus, the even and odd modes do not couple to each other. Moreover, the two passbands are independent. This characteristic can be utilized in the design of dual-band BPF with independently two passbands.

Figure 2.9 shows simulated S_{21}-magnitude of a weak coupling resonator circuit under different orders of odd–even-modes, where odd_b and even_b denote the odd and even mode provided by the bottom dual-mode resonator and odd_t and even_t denote the odd and even mode provided by the top dual-mode resonator as shown in Fig. 2.1. The orders of odd–even-modes are controlled by electrical length of θ_{4b} and θ_{4t}. It can be observed that Fig. 2.9a has the best selectivity because of one TZ located at lower first passband, one TZ located between two passbands and one TZ located at higher second passband. Figure 2.9b shows a better selectivity between first and second passband. However, poor selectivity exists at the low frequency. There are only two TZs in the desired band, which resulting in poor selectivity, as shown in Fig. 2.9c, d.

By placing a short or open circuit at the symmetric plane of the circuit as shown in Fig. 2.2, we obtain the circuit model for even or odd modes as shown in Fig. 2.10a, c. To demonstrate how to control these modes, the circuit models in Fig. 2.10a, c can be modified as those in Fig. 2.10b, d, respectively. The resonant frequencies can be calculated when $\mathrm{Im}[Y_{even}] = 0$ and $\mathrm{Im}[Y_{odd}] = 0$ from the end of the even- and odd-mode circuits, respectively, which can be expressed by for even mode

$$Y_2 \frac{(Y_{Load} + Y_L) + jY_2 \tan(\beta L_2)}{Y_2 + j(Y_{Load} + Y_L)\tan(\beta L_2)} + jY_1 \tan(\beta L_1) = 0 \tag{2.5}$$

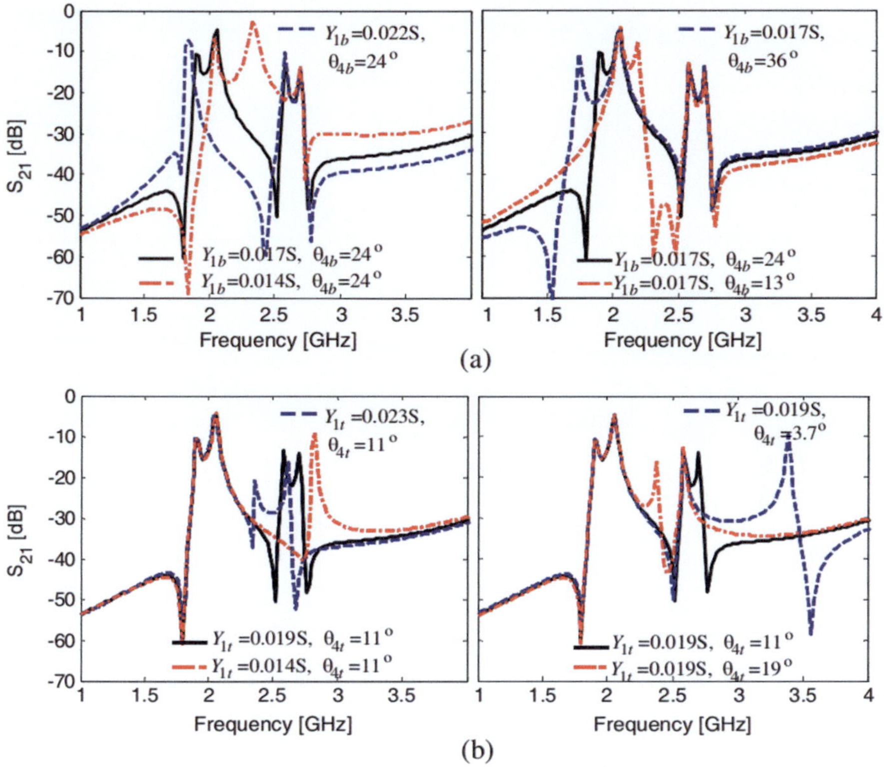

Fig. 2.8 Simulated S_{21}-magnitude of a weak coupling resonator circuit. **a** Resonant frequency according to Y_{1b} and θ_{4b}. **b** Resonant frequency according to Y_{1t} and θ_{4t}

where

$$Y_L = Y_3 \frac{jY_4 \tan(\beta L_4) + jY_3 \tan(\beta L_3)}{Y_3 - Y_4 \tan(\beta L_3)\tan(\beta L_4)} \tag{2.6}$$

Y_{Load} represents the input impedance from the via to the CPW portion. Since the capacitances are small for metallic via, Y_{Load} can be estimated from

$$Y_{Load} = \frac{j\frac{Y_2}{2}\tan\left(\beta l_5 + \frac{\beta L_6}{2}\right) + 1}{1 - Y_2 \omega L \tan\left(\beta L_5 + \frac{\beta L_6}{2}\right)} \tag{2.7}$$

where the β and ω are the propagation constant and frequency of the electromagnetic waves in the medium.

For odd mode, the resonant frequencies can be calculated as

$$-jY_2 \cot(\beta L_2) + jY_1 \tan(\beta L_1) = 0 \tag{2.8}$$

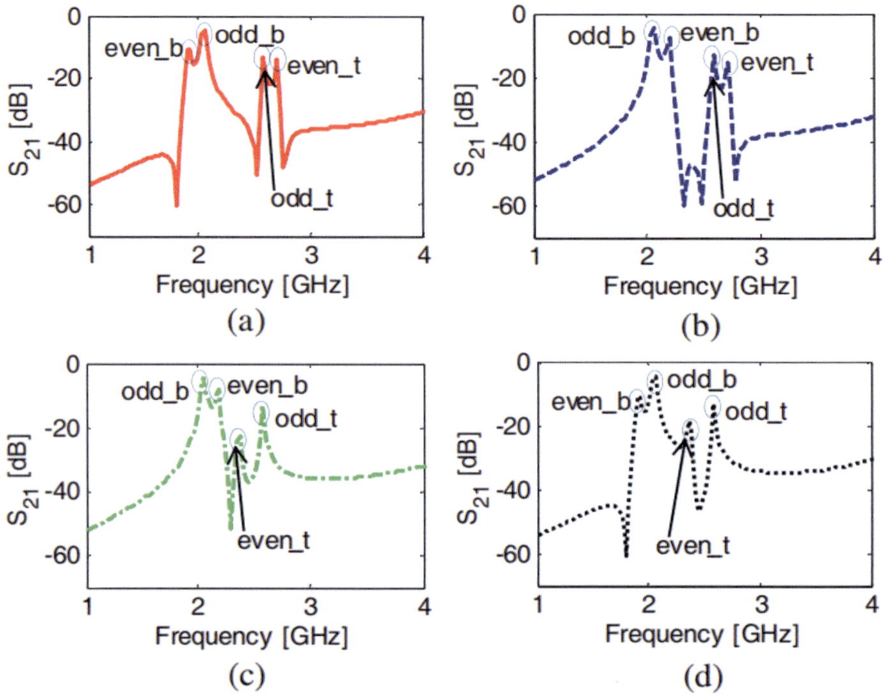

Fig. 2.9 S_{21}-magnitude under different orders of odd–even-modes. **a** $\theta_{4b} = 24°$, $\theta_{4t} = 11°$. **b** $\theta_{4b} = 13°$, $\theta_{4t} = 11°$. **c** $\theta_{4b} = 13°$, $\theta_{4t} = 4°$. **d** $\theta_{4b} = 24°$, $\theta_{4t} = 4°$

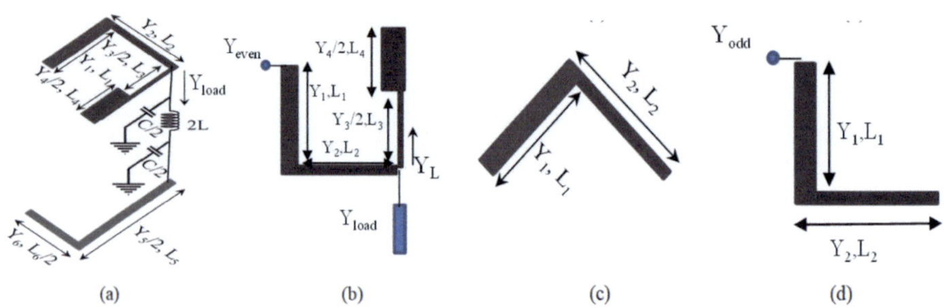

Fig. 2.10 **a**, **b** Even mode equivalent circuit models. **c**, **d** Odd mode equivalent circuit models

It can be seen from (2.5) and (2.8) that the odd mode does not vary so much with microstrip open stub and CPW structure while the even-mode frequencies can be tuned by microstrip open stub and CPW structure. Figure 2.11 shows the simulated resonance properties of the proposed resonator against Y_4 and the length L_6 under weak coupling. The other dimensions are $Y_1 = 0.0167$ S, $Y_2 = 0.0125$ S, $Y_3 = 0.0145$ S, $Y_5 = 0.0167$ S, $Y_6 = 0.0263$ S, $L_1 = 9.4$ mm, $L_2 = 11.2$ mm, $L_3 = 8.65$ mm, $L_4 = 6.5$ mm and $L_5 = 16$ mm, the diameter of the via is 0.8 mm. It can

Fig. 2.11 Resonant frequencies versus Y_4 and L_6

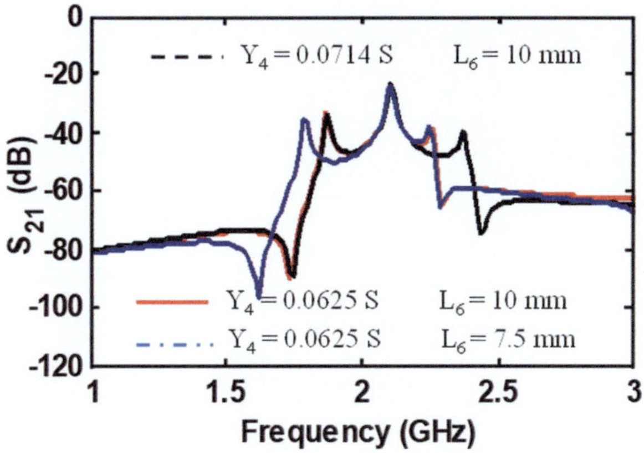

be observed that there are three resonant frequencies in the desired band: two even modes and one odd mode frequency. As the admittance Y_4 varies from 0.0625 to 0.0714 S, the lower even-mode resonant frequency moves down while the odd mode and other even modes remain the same. When the L_6 decreases from 1.00 to 0.75 mm, the higher even-mode resonant frequency moves up while the odd mode and other even-modes keep unchanged. Thus, the two even modes and one odd mode can be independently tuned to meet the requirement of the passband's characteristics. Two transmission zeros are generated at both sides of the even-mode resonance frequency due to the characteristics of odd–even mode and variation in the even mode resonance frequency.

2.3.2 *Theoretical Analysis of BPF Based on SIW Multi-mode Resonators*

To achieve a magnetic coupling based on SIW, we use the Filter IV structure shown in Fig. 2.5. Two slots are placed at the sidewalls on the middle metal layer, where one can not only have the maximum magnetic field of TE_{101} mode but also the minimum field distribution of TE_{102} mode, thanks to the orthogonality of those Eigen-resonant modes. The slots enable the two resonators coupled magnetically while minimizing the TE_{102} Coupling.

A set of simulated results are plotted in Fig. 2.12a. The coupling coefficients of TE_{101} and TE_{102} mode can both be extracted by

$$k = \pm \frac{f_1^2 - f_2^2}{f_1^2 + f_2^2} \tag{2.9}$$

On one hand, the coupling coefficients between TE_{101} modes are magnetic and can be tuned by width W_s of the two slots. On the other hand, the coupling coefficients

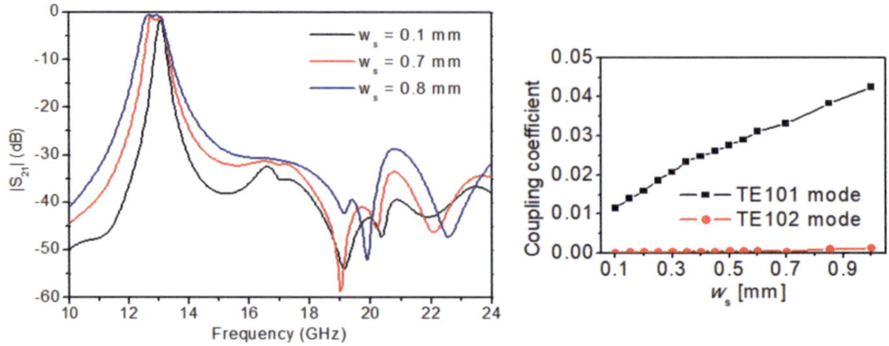

Fig. 2.12 Extracted coupling coefficients of TE_{101} and TE_{102} and simulated $|S_{21}|$ against w_s

of TE_{102} mode increase slightly, still keeping a negligible value. This phenomenon is also verified by the simulated S-parameters of Filter IV as shown in Fig. 2.12b. In line with the increasing width of slots, the filter bandwidth is also getting larger. In parallel, the first harmonic generated by TE_{102} mode becomes more visible even through it is still small enough (below -30 dB). Based on the above analysis, Filter IV can preserve the magnetic coupling while suppressing the first higher order mode TE_{102}.

In Filter V, a circular aperture with radius E_{r1} is placed at the center on the middle metal layer, where the maximum electric field takes place to enable the electric coupling of TE_{101} mode. On the other hand, TE_{102} mode is also magnetically coupled through the same aperture. To counter-balance the unnecessary or excessive magnetic coupling of TE_{102} mode, a mixed coupling scheme is proposed. To do so, two small circular apertures with radius E_{r2} are placed at the middle metal layer, which exhibit the maximum electrical field of TE_{102} mode, as shown in the layout of Filter V in Fig. 2.5. The overall coupling coefficient can be expressed by

$$k_{all} = |k_m - k_e| \tag{2.10}$$

The bandwidth of the passband can be controlled by varying E_{r1}. Then, by adjusting E_{r2}, the overall coupling coefficients of TE_{102} mode can be magnetic, zero or electric, respectively. The extracted results using (2.9) and (2.10) are presented in Fig. 2.13. As indicated in Fig. 2.13a, the overall coefficient of TE_{101} mode can be tuned by adjusting E_{r1} while preserving a constant E_{r2}. The bandwidth of Filter V is getting larger while increasing E_{r1}. Then, in Fig. 2.13b, along with the increasing of radius E_{r2} from 0 mm, the overall coupling coefficients of TE_{102} mode decreases. The coefficient is quite close to zero for $E_{r2} = 0.79$ mm (less than 0.002). As E_{r2} continuously increases, the overall coefficient of TE_{102} starts to increase from almost zero value. To valid the extracted results, the simulated results of $|S_{21}|$ of Filter V are also presented in Fig. 2.13b. A harmonic passband is generated at the resonant frequency of TE_{102} mode without any suppression, as the solid black line shown in Fig. 2.13b.

Furthermore, when $E_{r2} \leq 0.78$ mm, the mixed-coupling becomes magnetic, a transmission zero is then generated at the low-side of the resonant frequency of TE_{102}. When $E_{r2} = 0.79$ mm, the overall coupling coefficient of TE_{102} mode is almost zero so that the harmonic produced by TE_{102} is suppressed below $-$ 20 dB. When $E_{r2} \geq 0.80$ mm, the mixed-coupling is electric, which introduces a transmission zero at the up-side of the resonant frequency of TE_{102} mode. At the same time, it can be seen from the simulated $|S_{21}|$ and extracted coupling coefficients that the tuning of E_{r2} introduces a limited influence on the passband related to the fundamental TE_{101} modes. The increase of E_{r2} only introduces a weak electric coupling of TE_{101}, which has a little influence on the overall coupling of the fundamental mode. The simulation of $|S_{21}|$ also coincides with the extracted coupling coefficients presented in Fig. 2.13.

Through the mixed-coupling scheme of higher-order mode TE_{102}, its harmonic can be suppressed. Furthermore, the coupling between TE_{101} modes can still keep electric.

Next, we focus on the core theoretical analysis of four-band multilayer SIW filter: Filter VI. In order to decide the center frequencies of the passbands and avoid the

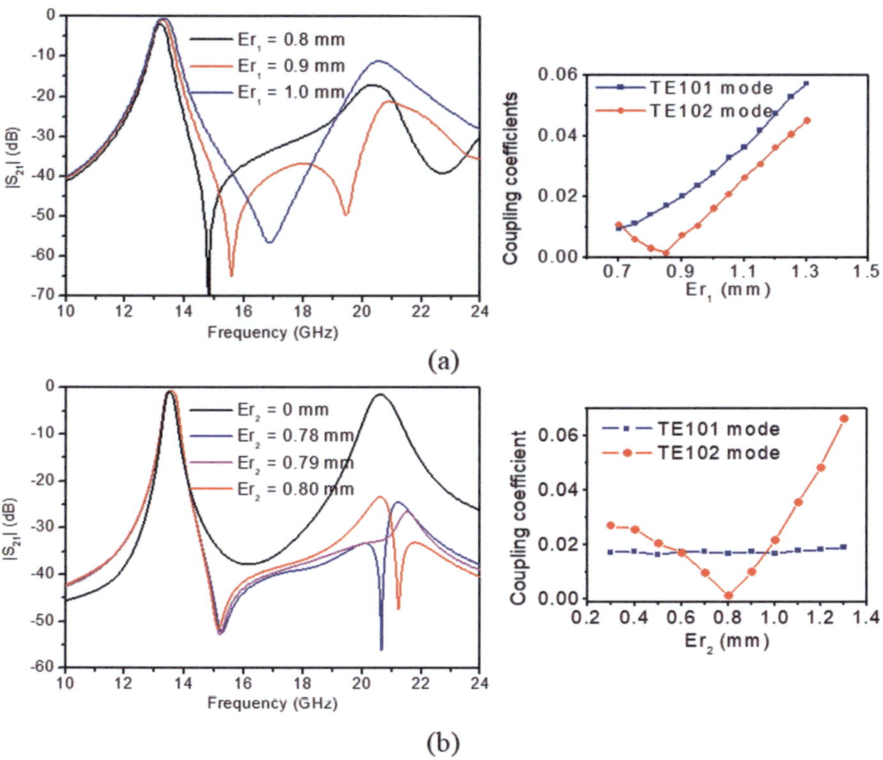

Fig. 2.13 Extracted coupling coefficients of TE_{101} and TE_{102} modes and simulated $|S_{21}|$. **a** Against E_{r1} ($E_{r2} = 0.8$ mm). **b** Against E_{r2} ($E_{r1} = 0.9$ mm)

unwanted modes allocated in the frequency range of the four passbands, the dimensions of the SIW resonator should be accurately chosen. The resonant frequencies of TE_{m0n} mode can be calculated by the equations presented in [37] as

$$f_{TE_{m0n}} = \frac{c}{2\pi\sqrt{\varepsilon_r}}\sqrt{\left(\frac{m\pi}{a}\right)^2 + \left(\frac{n\pi}{l}\right)^2} \tag{2.11}$$

where a and l are the effective width and length of the SIW resonator. ε_r is the relative dielectric constant of the substrate, where m and n is the mode number of TE_{m0n} mode.

Defining the mode ratio as

$$r_{mn} = \frac{f_{TE_{m0n}}}{f_{TE_{101}}} \tag{2.12}$$

The ratio r_e of effective length l_{eff} and width w_{eff} of a SIW resonator is defined as:

$$r_e = \frac{l_{eff}}{w_{eff}} \tag{2.13}$$

Figure 2.14 presents the calculated results that are derived from (2.11), (2.12) and (2.13), considering all the modes near the passbands. The results are calculated under the conditions of variable l_{eff} with constant $w_{eff} = 9$ mm. When $r_e < 1.0$, the first four resonant modes of the rectangle SIW resonator are intended to be TE_{101}, TE_{201}, TE_{102} and TE_{202}. The maximum electric and magnetic fields of TE_{202} mode are completely overlapped with the maximum ones of TE_{201} and TE_{102}, which increases the difficulty to design a corresponding independent coupling of these modes. TE_{202} mode is therefore not suitable in our quad-band filter design. To avoid TE_{202}, r_e should be larger than 1.6. Furthermore, additional structure or unbalanced feed are required to perturb the two degenerate modes TE_{201} and TE_{102}, which introduces new design difficulty [38, 39]. In summary, TE_{10n} ($n = 1$, 2, 3 and 4) modes are more suitable to design the proposed quad-band filters, because the four modes can be generated by the same feeding and they have different distributions of maximum electromagnetic fields as possible.

In order to ensure that the passbands are formed by TE_{10n} modes, r_e should be larger than 2.2. The first, second and third higher-mode are TE_{102}, TE_{103} and TE_{104}, respectively. It also can be seen that r_{mn} decreases with the increase of r_e, which means that the frequency gap between different passbands of the filter becomes smaller. In other words, the center frequencies and its ratios of the four passbands of the proposed quad-band filter can be decided by the effective resonator width, length and its ratio r_e.

In a multi-mode multi-bands filter design, it is difficult to design independent couplings for different modes, which decides the bandwidths of the passbands. Thanks to the mode orthogonality of the resonant modes and the vertical coupled structure, it provides a relative independent coupling design. Figure 2.15 illustrates

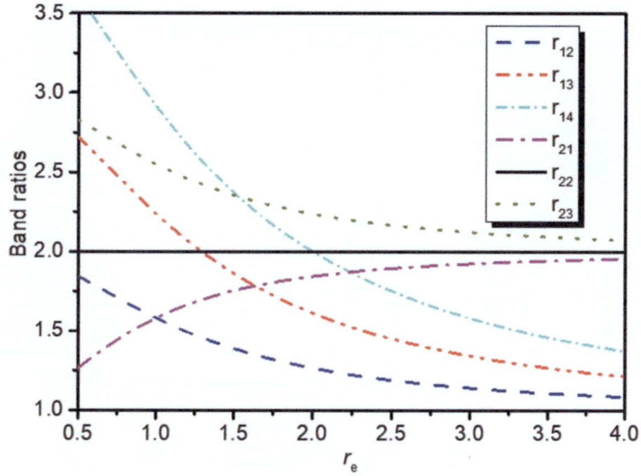

Fig. 2.14 Band ratios r_{mn} tuning against r_e

the field distribution of the selected four modes and its coupling scheme through the middle metal layer. c_n and s_m ($n = 1, 2, 3$ and $m = 1, 2, 3, 4$) represent the etched three circular apertures and four slots, respectively.

Firstly, the magnetic and electric fields of fundamental mode TE_{101} couple through s1, s2 and c1, respectively. Secondly, the magnetic field of TE_{102} mode couples through c1 while the electric field couples through c2 and c3. Thirdly, the electric field of TE_{103} mode couples through c1, c2 and c3, and the magnetic field couples through c2, c3, s3 and s4. Furthermore, the electric field of TE_{104} mode couples through s4. Since the width of s4 is small, the electric coupling of TE_{104} is small enough to be ignored. Meanwhile, the magnetic field of TE_{104} mode couples through c1, c2, c3 and s3. It can be seen that the four modes couple through different apertures, which can provide a relative independent method to control the couplings. Finally, the source-load coupling is introduced to produce more transmission zeros so as to improve the selectivity of the passbands.

To validate the coupling scheme, Fig. 2.16 illustrates how the bandwidth tuned by varying the parameters of the etched apertures. Thanks to the multi-mode effects and source-load coupling, two transmission zeros are allocated at the low- and high-side of each passband, respectively. In addition, one more transmission zero is generated at 11.5 GHz due to source-load coupling. As depicted in Fig. 2.16a, the bandwidth of first passband can be tuned by varying the width of s1 and s2 while the other passbands almost keep constant. This phenomenon can be explained by the coupling scheme presented in Fig. 2.15. For higher-order modes TE_{103} and TE_{104}, the maximum magnetic fields are mainly generated at the middle of the resonator, that is to say s1 and s2 mainly enable the magnetic couplings of TE_{101} mode and produce small influence on the other couplings. Similarly, the bandwidth of the second passband can be tuned by varying the radius of c2 and c3 without producing any influence on the first passband. Then, the bandwidth of the third passband can be tuned by varying the length of s3. Meanwhile, the first two passbands present too small changes that can be ignored. Finally, the fourth passband can be changed

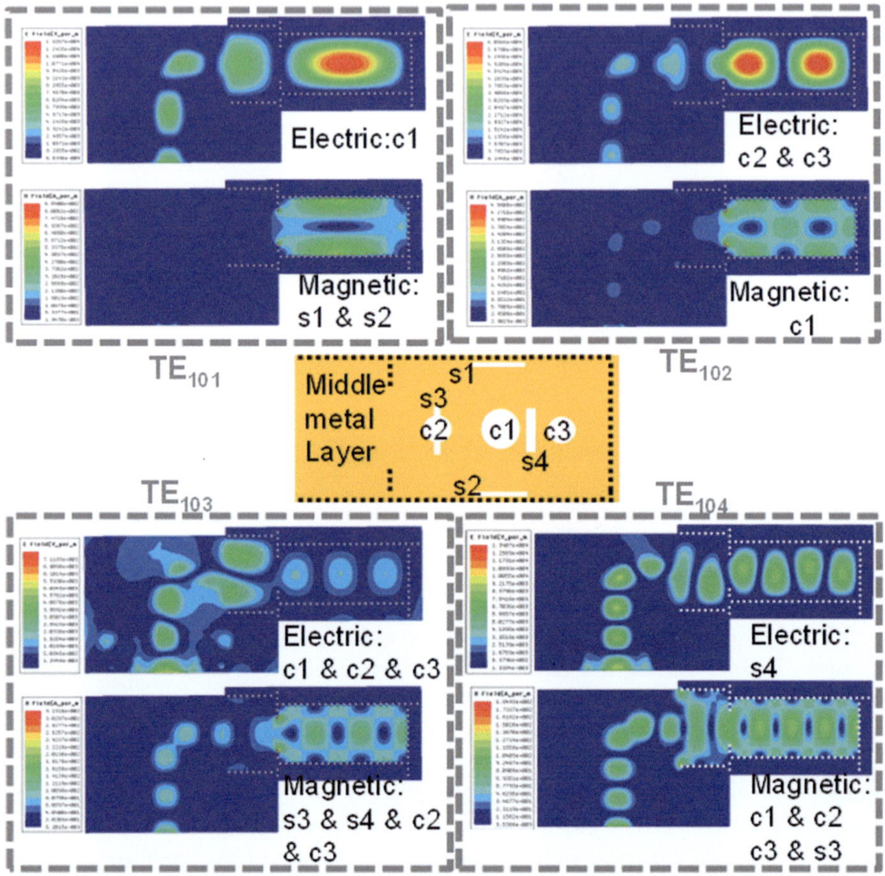

Fig. 2.15 Field distribution and coupling scheme of the selected four modes

by varying the length of s3. To keep the third passband unchanged, the length of s4, which only control the magnetic couplings of TE_{103} mode, needs to be tuned accordingly. By adopting the above-described methods, the bandwidths of the four passbands can be tuned independently. Additionally, the external factor Q_{ext} also can be used to determine the bandwidth, it is mainly controlled by the source-to-resonator coupling. In this work, the required Q_{ext} can be achieved by tuning w_1.

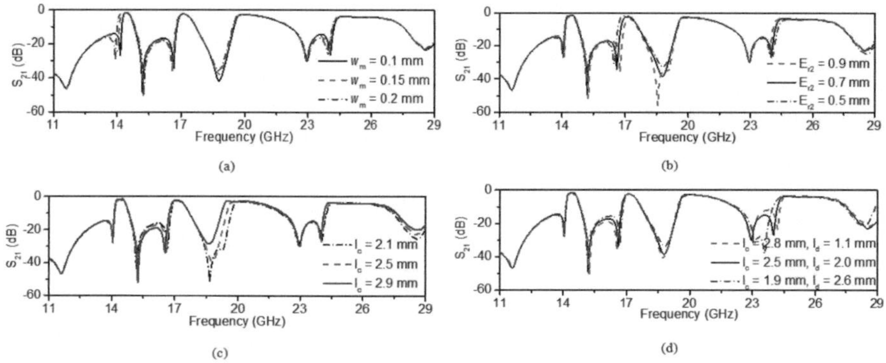

Fig. 2.16 Band tuning. **a** First passband. **b** Second passband. **c** Third passband. **d** Fourth passband

2.4 Validation of the Proposed High Selectivity Bandpass Filters

2.4.1 Planar Microstrip: Dual-Mode and Tri-Mode Resonators-Based Filters

Dual-Mode

To validate the aforementioned design scheme of a dual-passband high-selectivity filter based on dual-mode resonators, we fabricated and tested the dual-mode dual-band bandpass filter (Filter I) shown in Fig. 2.1. This filter was implemented on a 0.5 mm thick F4B-2 substrate (relative permittivity $\varepsilon_r = 2.65$) using microstrip etching technology, with overall dimensions of 44.2 mm × 30 mm. As shown in Fig. 2.17a, b, the measured results align well with simulation outcomes. At the two passbands centered at 2.0 and 2.65 GHz, the measured insertion loss is less than 0.6 dB for the first passband and below 0.9 dB for the second passband, while the return loss in both passbands exceeds 19 dB, indicating excellent passband performance. Particularly noteworthy is the clear presence of three TZs observed at 1.84, 2.46, and 2.81 GHz in the measured results. These zeros significantly enhance the filter's skirt selectivity: the 1.84 GHz TZ delivers strong rejection of 34 dB at the lower edge of the first passband; the 2.46 GHz TZ improves isolation between the two passbands; and the 2.81 GHz TZ provides deep rejection of 40 dB at the upper edge of the second passband. These experimental results convincingly demonstrate the feasibility of implementing dual-band filters with high selectivity and multiple transmission zeros based on dual-mode resonator structures.

Tri-Mode

Based on the research of dual-mode resonators, we designed and fabricated an improved filter (Filter II) as shown in Fig. 2.3, achieving a triple-mode resonator

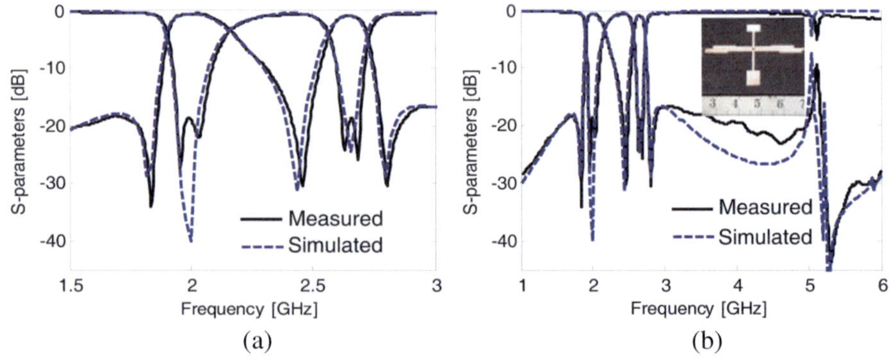

Fig. 2.17 Simulated, measured results and the photograph of the dual-band BPF: Filter I. **a** Narrowband response. **b** Wideband response

through a novel microstrip-CPW hybrid structure. This filter was also manufactured on a 0.5 mm thick F4B-2 substrate ($\varepsilon_r = 2.65$) and its photograph is presented in Fig. 2.18. Simulation and measured results in Fig. 2.19 demonstrate that while maintaining dual-passband characteristics at 2.0 and 2.65 GHz, the filter exhibits insertion loss below 0.6 dB in the first passband and under 1.1 dB in the second passband. The key improvement lies in introducing a fourth transmission zero through the CPW structure: in addition to retaining the original zeros at 1.8 GHz (34 dB rejection) and 2.82 GHz (40 dB rejection), two new zeros at 2.35 GHz (23 dB rejection) and 2.48 GHz (29 dB rejection) significantly enhance the transition steepness at the upper edge of the first passband and the lower edge of the second passband. The compact layout (27.0 mm × 5.5 mm) synergizes with four transmission zeros to achieve high-performance filtering with better than 23 dB inter-band isolation. Measured data strongly validates the outstanding capability of the microstrip-CPW hybrid resonator in enhancing passband selectivity and suppressing out-of-band interference.

We further designed, fabricated, and tested the triple-mode dual-band bandpass filter shown in Fig. 2.4, based on a microstrip-CPW structure. Implemented on

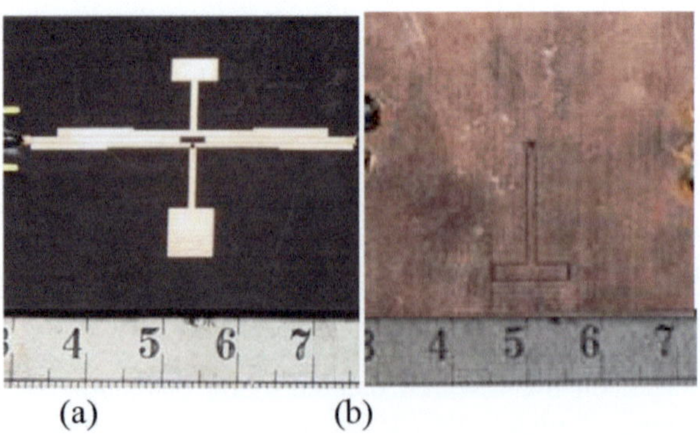

Fig. 2.18 Photograph of the fabricated dual-band BPF: Filter II. **a** Top view. **b** Bottom view

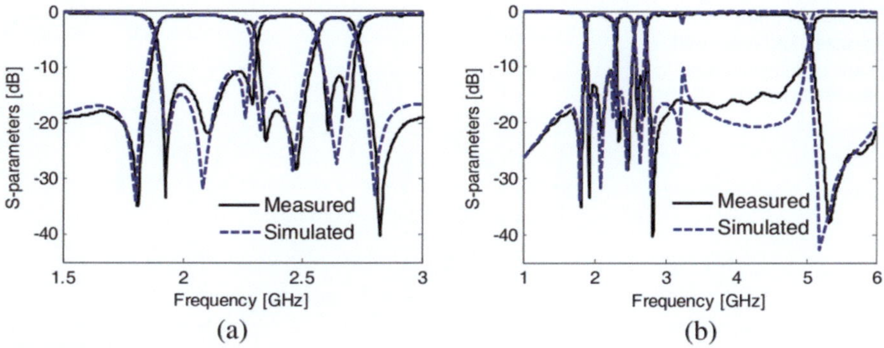

Fig. 2.19 Simulated, measured results and the photograph of the dual-band BPF: Filter II. **a** Narrowband and **b** wideband

0.5 mm F4B-2 substrate ($\varepsilon_r = 2.65$) with optimized dimensions in Table 2.1, its compact 25.2 mm × 30.0 mm footprint corresponds to merely $0.18\lambda_0 \times 0.21\lambda_0$ at 2.1 GHz. Full S-parameter characterization using an Agilent E5071C VNA shows strong correlation between measured and simulated results as shown in Fig. 2.20. Measurements confirm two distinct passbands centered at 2.1 GHz (15.2% 3 dB bandwidth) and 3.9 GHz (5% 3 dB bandwidth), with exceptional in-band performance: minimum insertion losses below 0.7 dB and 1.7 dB respectively, while return losses exceed 11 dB and 20 dB. The triple-mode resonator and coupling configuration efficiently generate five transmission zeros at 1.79 GHz (37 dB rejection), 2.27 GHz (19 dB), 3.49 GHz (47 dB), 3.80 GHz (33 dB), and 4.12 GHz (35 dB). Four near-passband-edge zeros originate from the resonator's inherent even/odd-mode properties and tunable resonant frequencies, while the inter-passband zero results from specific coupling mechanisms. Though measured insertion losses slightly exceed simulations due to fabrication tolerances, dielectric losses, and connector effects, overall correlation remains strong. These results successfully validate the proposed microstrip-CPW hybrid triple-mode design concept, achieving dual-band operation with high selectivity, low loss, and compact dimensions.

Table 2.1 Dimensions of tri-mode DBBPF (unit: mm)

Top	W_1	W_2	W_3	W_4	W_5	W_6	W_7	W_8	W_9	S_1
	0.6	1.0	0.8	6.0	0.6	0.9	0.8	4.0	21.7	0.2
	L_1	L_2	L_3	L_4	L_5	L_6	L_7	L_8	L_9	S_2
	11.2	9.4	8.65	6.5	5.1	6.1	4.2	4.1	0.55	0.15
Bottom	W_{10}	W_{11}	W_{12}	W_{13}	S_3	S_5				
	1.0	2.0	1.0	2.0	0.2	0.6				
	L_{10}	L_{11}	L_{12}	L_{13}	S_4	D_1				
	16.0	10.4	8.6	5.2	0.2	0.8				

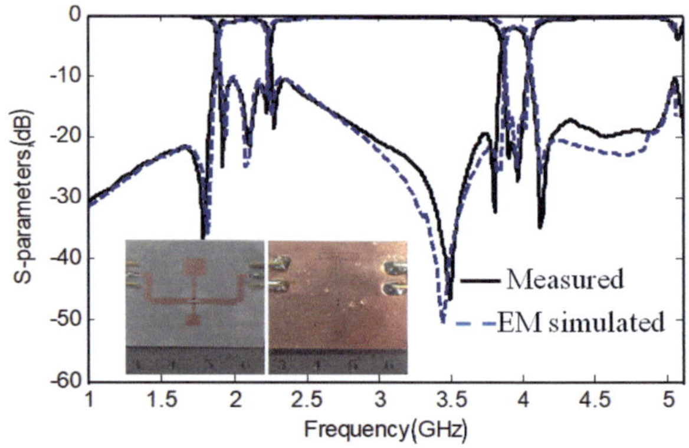

Fig. 2.20 Photograph, simulated and measured results of the proposed DBBPF: Filter III

2.4.2 Substrate-Integrated Waveguide: Multimode Resonators-Based Filters

To verify the proposed high-selectivity design approach for multi-layer SIW filters, we designed and fabricated two dual-layer structure filters (Filter IV and Filter V) based on different coupling mechanisms as shown in Fig. 2.5. Both filters were implemented on Rogers RT/Duroid 6002 substrate ($\varepsilon_r = 2.94$, thickness 0.762 mm), featuring magnetic coupling (Filter IV) and electric coupling (Filter V) structures achieved by etching specific apertures in the middle metal layer. Critical dimensions were optimized: resonator length $I_r = 9.1$ mm, magnetic coupling aperture width $W_s = 0.6$ mm (Filter IV), primary electric coupling aperture radius $E_{r1} = 0.9$ mm, and secondary aperture radius $E_{r2} = 0.8$ mm (Filter V). All structures achieved interlayer alignment through precision CNC milling, with RF ports employing 50 Ω microstrip-to-SIW tapered feeding structures. As shown in Fig. 2.21, measurements confirm both filters successfully suppressed harmonic passbands caused by the first higher-order TE_{102} mode; Filter IV suppressed spurious response at 20.5 GHz below − 30 dB with 1.7 dB insertion loss at its 13.0 GHz center frequency, while Filter V achieved − 20 dB suppression at the same frequency with reduced 1.5 dB insertion loss at 13.2 GHz. Both filters exhibit a 3 dB absolute bandwidth of 600 MHz and in-band return loss better than 15 dB. Notably, Filter V maintains over 20 dB rejection across a 15–30 GHz wide stopband due to its hybrid coupling design, significantly extending stopband performance. Experiments simultaneously verified the compact size advantage, with Filter V measuring just 9.1 mm × 5.0 mm, fully demonstrating the effectiveness of the proposed multilayer structures in achieving high selectivity, broadband suppression, and miniaturization.

To validate the effectiveness of the proposed quad-band SIW filter design methodology, we implemented the multilayer filter prototype shown in Fig. 2.6 based on $TE_{101}/TE_{102}/TE_{103}/TE_{104}$ multimode resonant mechanisms. Fabricated on Rogers RT/Duroid 6002 substrate ($\varepsilon_r = 2.65$, thickness 0.762 mm, $\tan\delta = 0.0012$) via photolithographic copper etching, this filter integrates input/output feed microstrips

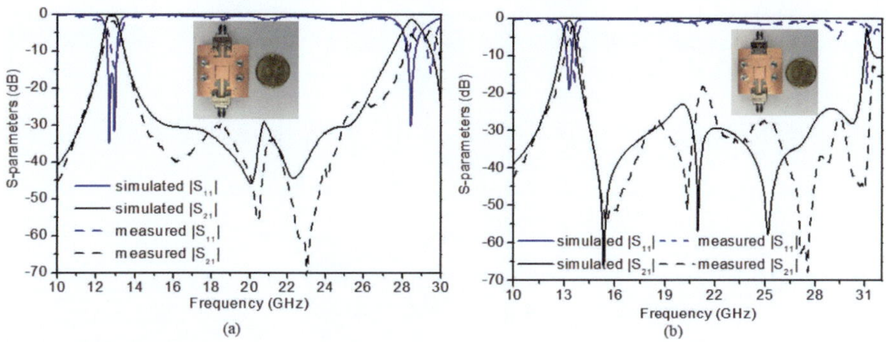

Fig. 2.21 Photographs, simulation and measured results of the proposed filters. **a** Filter IV (magnetic coupling). **b** Filter V (electric coupling)

on the top and bottom metal layers, with selectively etched coupling windows in the middle layer enabling mode-selective coupling. Laser alignment technology ensured ± 5 μm interlayer registration precision during assembly, and SMA-K connectors were integrated at feed ports to minimize measurement errors.

Measured S-parameters in Fig. 2.22 demonstrate four passbands centered at 14.4, 17.0, 20.4 and 25.5 GHz with less than 0.3% deviation from theoretical values. The first passband achieved 1.6 dB insertion loss within a 300 MHz absolute bandwidth (unloaded Q $\approx$ 420), flanked by transmission zeros at 13.8 and 15.1 GHz providing over 45 dB out-of-band rejection. Notably, the second passband maintained just 2.2 dB loss across 520 MHz bandwidth while generating three transmission zeros at 16.2/17.8/18.5 GHz through s_3/s_4 slot combination control, enabling exceptional 120 dB/GHz roll-off characteristics. Employing hybrid electromagnetic coupling, the third and fourth passbands preserved low insertion losses of 1.3 dB/2.5 dB across wide bandwidths of 1640 MHz/2660 MHz, with zeros at 24.1 and 26.8 GHz effectively suppressing high-frequency spurious responses. Return losses exceeded 10 dB across all bands with less than 1% center frequency drift, confirming the vertical coupling structure's independent control over each passband.

Compared to conventional single-layer SIW filters, this design achieves breakthrough performance within a compact 915 mm^2 footprint: (1) nine transmission zeros generated at passband edges through source-load coupling, delivering typical > 40 dB out-of-band rejection; (2) independent tunability of absolute bandwidths; (3) low 2.5 dB insertion loss maintained at millimeter-wave frequencies (25.5 GHz), verifying effective high-frequency loss suppression. Measurement-to-simulation correlation exceeded 90% with frequency tolerance within ± 1.5%, fully validating the engineering practicality of the proposed vertical coupling mechanism and multimode co-design theory.

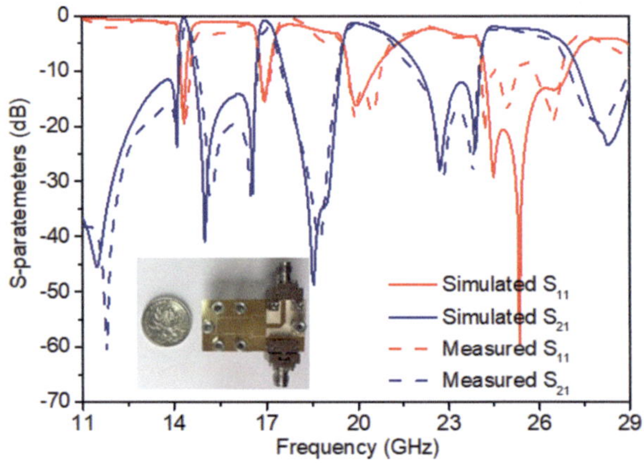

Fig. 2.22 Photograph, simulated and measured results of the proposed filter: Filter VI

2.5 Conclusion

This chapter systematically explores the design methodology of multi-band, high-selectivity bandpass filters based on multi-mode resonators, with its core innovations crucially lying in resonator structure reconfiguration, coupling mechanism innovation, and multi-mode control strategies. In the planar microstrip domain, a dual independent tunable dual-mode resonator structure utilizing common port feeding was first proposed, achieving independent dual-passband control while effectively generating TZs that enhance band-edge selectivity. To further improve the high-end selectivity of the first passband, a three-dimensional hybrid microstrip and CPW structure was designed. By introducing an additional third even-mode resonance point, this structure successfully transformed the dual-mode resonator into a tri-mode resonator within the same physical footprint. Building upon this, a composite tri-mode resonator architecture was proposed, vertically integrating microstrip open stubs with CPW resonant units on the ground layer through metal vias. This implementation achieved precise control over five TZs, significantly enhancing the roll-off characteristics of the dual passbands and inter-band isolation. To overcome the limitations of planar structures in multi-band expansion and independent mode control, the design concept was extended to multi-layered SIW technology. Compact dual-band filter designs based on vertical coupling and resonant suppression were respectively presented, with a key breakthrough involving the use of a vertically stacked multi-layer coupling architecture: precisely suppressing higher-order modes through magnetic coupling, controllable electric coupling, and hybrid coupling cancellation mechanisms realized by specific apertures in the intermediate metal layer. Finally, capitalizing on the full utilization of orthogonally distributed TE_{m0n} modes and the advantages of three-dimensional coupling structures, a quad-band SIW filter with independent bandwidth control capability was designed, enabling relatively independent tuning of the bandwidths for its four distinct passbands. Through theoretical modeling, parameter equation derivation, and comprehensive simulation and measurement validation, the effectiveness and superiority of the proposed design schemes were thoroughly demonstrated in realizing high-selectivity, low-loss, miniaturized multi-band filters.

References

1. Chen, C. Y. and C. C. Lin, "The design and fabrication of a highly compact microstrip dual-band bandpass filter," Progress In Electromagnetics Research, Vol. 112, 299–307, 2011.
2. Velazquez-Ahumada, M. D. C., "Application of stub loaded folded stepped impedance resonators to dual band filter design," Progress In Electromagnetics Research, Vol. 102, 107–124, 2010.
3. Kuo, J. T. and S. W. Lai, "New dual-band bandpass filter with wide upper rejection band," Progress In Electromagnetics Research, Vol. 123, 371–384, 2012.
4. Wu, H. W., Y. F. Chen, and Y. F. Chen, "Multi-layered dual-band bandpass filter using stub-loaded stepped-impedance and uniform-impedance resonators," IEEE Microw. Wireless Compon. Lett., Vol. 22, No. 3, 114–116, Mar. 2012.
5. Wang, J., L. Ge, K. Wang, and W. Wu, "Compact microstrip dual-mode dual-band bandpass filter with wide stopband," Electronics Letters, Vol. 47, No. 4, Feb. 2011.
6. Chen, C. F., T. Y. Huang, and R. B. Wu, "Design of dual- and triple-passband filters using alternately cascaded multiband resonators," IEEE Trans. on Microw. Theory and Tech., Vol. 54, No. 9, 3550–3558, Sep. 2006.
7. Deng, P. H. and J. H. Jheng, "A switched reconfigurable high-isolation dual-band bandpass filter," IEEE Microw. Wireless Compon. Lett., Vol. 21, No. 2, 71–73, Feb. 2012.
8. Xu, J., C. Miao, and W. Wu, "A compact and high isolation dual-mode dual-band bandpass filter with tunable transmission zeros," Journal of Electromagnetic Waves and Applications, Vol. 26, Nos. 17–18, 2390–2397, Dec. 2012.
9. Wei, F., L. Chen, and X. W. Shi, "Compact dual-mode dual-band bandpass filter with wide stopband," Journal of Electromagnetic Waves and Applications, Vol. 26, Nos. 11–12, 1441–1447, Aug. 2012.
10. Deng, H. W., Y. J. Zhao, X. S. Zhang, W. Chen, and J. K. Wang, "Compact and high selectivtity dual-band dual-mode microstrip BPF with single step-impedance resonator," Electronics Letters, Vol. 47, No. 5, Mar. 2011.
11. Youngje, S., "Dual-mode dual-band filter with band notch structures," IEEE Microw. Wireless Compon. Lett., Vol. 20, No. 2, 73–75, Jan. 2010.
12. Sun, S., "A dual-band bandpass filter using a single dual-mode ring resonator," IEEE Microw. Wireless Compon. Lett., Vol. 21, No. 6, 298–300, Jun. 2011.
13. Lee, J. and Y. Lim, "Compact dual-band bandpass filter with good frequency selectivity," Electronics Letters, Vol. 47, No. 25, Dec. 2011.
14. Wu, Y. L., C. Liao, and X. Z. Xiong, "A dual-wideband bandpass filter based on E-shaped microstrip SIR with improved upper stopband performance," Progress In Electromagnetics Research, Vol. 108, 141–153, 2010.
15. Chang, W. S. and C. Y. Chang, "Analytical design of microstrip short-circuit terminated stepped-impedance resonator dual-band filters," IEEE Trans. on Microw. Theory and Tech., Vol. 59, No. 7, 1730–1739, Jul. 2011.
16. Mashhadi, M. and N. Komjani, "Design of novel dual-band bandpass filter with multi-spurious suppression for WLAN application," Journal of Electromagnetic Waves and Applications, Vol. 26, No. 7, 851–862, 2012.
17. Seungku, L. and L. Yongshik, "A planar dual-band filter based on reduced-length parallel coupled lines," IEEE Microw. Wireless Compon. Lett., Vol. 20, No. 1, 16–18, Jan. 2010.

18. D. Deslandes and K. Wu, "Single-substrate integration technique of planar circuits and waveguide filters," IEEE Trans. Microw. Theory Tech., vol. 51, no. 2, pp. 593–596, Feb. 2003.
19. S. S. Xu, K. X. Ma, F. Y. Meng and K. S. Yeo, "Novel defected ground structure and two-side loading scheme for miniaturized dual-band SIW bandpass filter designs," IEEE Microw. Wireless Compon. Lett., vol. 25, no. 4, pp. 217–219, Feb. 2015.
20. D. D. Lv, L. Q. Meng and Z. Zou, "Miniaturized HMSIW dual-band filter based on CSRRs and microstrip open-stubs," Prog. Electromagn. Res. Lett., vol. 77, pp. 97–102, July. 2018.
21. M. Esmaeili and J. Bornemann, "Substrate integrated waveguide triple-passband dual-stopband filter using six cascaded singlets," IEEE Microw. Wireless Compon. Lett., vol. 24, no. 7, pp. 439–441, Jul. 2014.
22. X. P. Chen, K. Wu and Z. L. Li, "Dual-band and triple-band substrate integrated waveguide filters with Chebyshev and quasi-elliptic responses," IEEE Trans. Microw. Theory Tech., vol. 55, no. 12, pp. 2569–2578, Dec. 2007.
23. B. J. Chen, T. M. Shen and R. B. Wu, "Dual-Band vertically stacked laminated waveguide filter design in LTCC technology," IEEE Trans. Microw. Theory Tech., vol. 57, no. 6, pp. 1554–1562, Jun. 2009.
24. W. Shen, W. Y. Yin and X. W. Sun, "Miniaturized dual-band substrate integrated waveguide filter with controllable bandwidths," IEEE Microw. Wireless Compon. Lett., vol. 21, no. 8, pp. 418–420, Aug. 2011.
25. Y. D. Wu, G. H. Li, W. Yang and T. Mou, "A novel dual-band SIW filter with high selectivity," Prog. Electromagn. Res. Lett., vol. 60, pp. 81–88, May. 2016.
26. K. Zhou, C. X. Zhou and W. Wu, "Substrate-integrated waveguide dual-mode dual-band bandpass filters with widely controllable bandwidth ratios," IEEE Trans. Microw. Theory Tech., vol. 65, no. 10, pp. 3801–3812, Oct, 2017.
27. X. P. Chen, K. Wu and D. Drolet, "Substrate integrated waveguide filter with improved stopband performance for satellite ground terminal," IEEE Trans. Microw. Theory Tech., vol. 57, no. 3, pp. 674–683, Mar. 2009.
28. X. P. Chen and K. Wu, "Substrate integrated waveguide cross-coupled filter with negative coupling structure," IEEE Trans. Microw. Theory Tech., vol. 56, no. 1, pp. 142–149, Jan. 2008.
29. W. Shen, L. S. Wu, X. W. Sun, W. Y. Yin and J. F. Mao, "Novel substrate integrated waveguide filters with mixed cross coupling (MCC)," IEEE Microw. Wireless Compon. Lett., vol. 19, no. 11, pp. 701–703, Nov. 2009.
30. B. Lee, T. H. Lee, K. Lee, M. S. Uhm and J. Lee, "K-band substrate-integrated waveguide resonator filter with suppressed higher-order mode," IEEE Microw. Wireless Compon. Lett., vol. 25, no. 6, pp. 367–369, Jun. 2015.
31. R. Moro, S. Moscato, M. Bozzi and L. Perregrini, "Substrate integrated folded waveguide filter with out-of-band rejection controlled by resonant-mode suppression," IEEE Microw. Wireless Compon. Lett., vol. 25, no. 4, pp. 214–216, Apr. 2015.
32. K.-S. Chin, C.-C. Chang, C.-H. Chen, Z. X. Guo, D. S. Wang, and W. Q. Che, "LTCC multilayered substrate-integrated waveguide filter with enhanced frequency selectivity for system-in-package applications," IEEE Trans. Compon., Packag., Manuf. Technol., vol. 4, no. 4, pp. 664–672, Apr. 2014.
33. Xiao-Guo Huang, Quanyuan Feng, Qian-Yin Xiang, and Ding-Hong Jia, "A High Selectivity Dual-Band Bandpass Filter Using Dual-Mode and Triple-Mode Resonators," Progress In Electromagnetics Research C, Vol. 36, 81–90, 2013. https://doi.org/10.2528/PIERC12121701.
34. X. Huang, Q. Feng, L. Zhu, Q. Xiang and D. Jia, "Triple-mode dual-band bandpass filter on microstrip-to-CPW structure," 2014 IEEE International Wireless Symposium (IWS 2014), Xi'an, China, 2014, pp. 1–4, https://doi.org/10.1109/IEEE-IWS.2014.6864193.
35. D. Jia, Q. Feng, Q. Xiang and K. Wu, "Multilayer Substrate Integrated Waveguide (SIW) Filters With Higher-Order Mode Suppression," in IEEE Microwave and Wireless Components Letters, vol. 26, no. 9, pp. 678–680, Sept. 2016, https://doi.org/10.1109/LMWC.2016.2597222.
36. Ding-Hong Jia, Jianqin Deng, Yangping Zhao, and Ke Wu, "Quad-Band Multilayer SIW Filter with High Selectivity and Controllable Bandwidths," Progress In Electromagnetics Research Letters, Vol. 83, 133–138, 2019. https://doi.org/10.2528/PIERL19021801.

37. D. M. Pozar, Microwave engineering, Third Edition. New York: Wiley, 2005.
38. C. C. Chuang, H. H. Lin and C. L. Wang, "Design of dual-mode SIW cavity filters," 2007 IEEE Region 10 Conference, Taipei, Taiwan, Oct, 2007.
39. M. Almalkawi, L. Zhu and V. Devabhaktuni, "Dual-mode substrate integrated waveguide (SIW) bandpass filters with an improved upper stopband performance," 2011 International Conference on Infrared, Millimeter, and Terahertz Waves, Houston, TX, USA, Oct, 2011.

Chapter 3
Tunable Bandpass Filters with Controllable Bandwidth

3.1 Introduction

Modern wireless communication systems, radar, spectrum sensing, and multi-standard transceivers impose increasingly stringent performance requirements on filters. Conventional tunable filters typically cannot independently adjust bandwidth and center frequency, resulting in mutual constraints between these parameters that often cause signal distortion or reduced spectral efficiency. For practical deployment, filters must not only possess frequency tuning capabilities to accommodate dynamic channel allocation and anti-jamming demands but also maintain consistent passband stability during tuning operations. This stability manifests primarily as three critical types: filters maintaining constant absolute bandwidth during center frequency shifts [1–9]; filters preserving fixed center frequencies while tuning passband bandwidth [10–12]; and filters enabling simultaneous tuning of both center frequency and bandwidth [13–16]. Such filters deliver significant value in reconfigurable wireless front-ends. Firstly, RF tunable filters intrinsically reduce overall size, complexity, and cost in multi-band systems [17–19]. Secondly, the constant bandwidth feature ensures stable spectral performance during frequency tuning, preventing bandwidth compression or expansion issues. Thirdly, the fixed center frequency with tunable bandwidth allows flexible adaptation to varying-bandwidth signals at single operating frequencies and enables real-time interference avoidance, substantially improving spectral efficiency and interference immunity.

Against this backdrop, researchers have conducted in-depth investigations into various tunable filters and proposed multiple solutions for independently adjustable bandwidth and center frequency. These approaches primarily fall into two technical categories: schemes based on independent electromagnetic coupling or hybrid electromagnetic coupling [1–6], and those achieved through specific resonator configurations. For instance, Ref. [7] successfully constructed a tunable filter with constant absolute bandwidth (CABW) characteristics using dual combline resonators combined with a lumped series resonator; Refs. [8, 9] developed a class of tunable

F. Nian et al., *RF & Microwave Multi-mode and Tunabe Filters*,
https://doi.org/10.1007/978-981-92-0577-6_3

filters by controlling the tuning rates of two resonant frequencies in dual-mode microstrip resonators to achieve CABW. Related research has further extended to higher-order tunable bandpass filters [10, 11, 13, 15, 20–22]. Reference [10] demonstrates a ring resonator bandpass filter with switchable bandwidth, where PIN diodes and stepped-impedance stubs are utilized for bandwidth adjustment. Reference [11] reports a varactor-tuned quadruple-mode bandpass filter with independently adjustable upper and lower bandwidths. Reference [13] proposes a dual-mode triangular patch bandpass filter achieving independent control of degenerate fundamental mode frequencies through two varactor-loaded slots, enabling simultaneous tuning of center frequency and bandwidth. Reference [15] proposed another tri-pole quasi-elliptic tunable filter that achieves comparable performance through adjacent coupling paths of combline resonators while improving roll-off characteristics via non-adjacent coupling paths. Reference [20] reported a tri-pole combline tunable filter with controllable bandwidth, where an external closed-loop circuit creates tunable transmission zeros in the low-frequency band to enhance selectivity. Reference [21] implemented a tri-pole tunable filter using a cross-shaped multi-mode resonator, enabling passband tunability through full control of resonant modes and external quality factor (Q_{ext}). Reference [22] presented a quad-pole elliptical combline filter with tunable center frequency and controllable bandwidth, significantly enhancing selectivity by generating and allocating a pair of tunable transmission zeros at both passband edges.

This chapter focuses on designing bandpass filters with independently tunable bandwidth and center frequency. The core mechanism for achieving constant absolute bandwidth lies in counteracting bandwidth variations during frequency tuning through electromagnetic independent/hybrid coupling combined with dynamic control of Q_{ext}. For example, co-directional open-loop resonators simultaneously excite electric and magnetic coupling at open-gap edges, enabling adaptive coupling coefficients with frequency, while tunable capacitors at input/output ports precisely regulate Q_{ext}—jointly ensuring bandwidth stability. Furthermore, solutions based on J/K inverter theory establish mathematical constraints between coupling strength and center frequency through precise modeling to implement high-order CABW filters. For bandwidth-tunable filters with fixed center frequency, the critical approach employs resonant mode decoupling techniques. One method utilizes multi-mode resonator structures to independently shift passband edges by differentiating physical control paths for odd- and even-modes. In a triple-mode resonator, the center frequency is fixed by the odd-mode frequency, while two even-modes controlled separately by stub varactors continuously tune the upper/lower band edges within a specific range. Another approach based on cross-coupling stabilizes coupling coefficients through optimized parallel-coupled microstrip parameters, determines Q_{ext} via capacitor-transmission line input networks, and dynamically adjusts transmission zero positions using independently controlled electric cross-coupling varactors to optimize passband roll-off characteristics while maintaining center frequency.

The chapter is organized as follows: Sect. 3.2 details several bandpass filter topologies with independently tunable bandwidth/center-frequency and their operating principles. Section 3.3 performs theoretical analysis to derive mathematical models and control equations for key parameters (e.g., coupling coefficient k, external quality factor Q_{ext}), clarifying conditions for constant-bandwidth and constant-center-frequency implementations. Section 3.4 presents comprehensive design examples with simulation and measured results to validate the effectiveness of proposed methods, topologies, and parametric theories. Finally, Sect. 3.5 summarizes the main findings and contributions.

3.2 Proposed Topology of Tunable Bandpass Filters with Controllable Bandwidth

In this chapter, we focus on several innovative topologies designed for bandwidth-controllable tunable bandpass filters. The core objectives include achieving constant absolute bandwidth (CABW) and constant fractional bandwidth (CFBW), as well as enabling dynamic, independent bandwidth tuning at fixed center frequencies.

To address the challenges of implementing constant absolute bandwidth in conventional tunable filters, we first propose a topology based on co-directional dual-open-loop resonators, as illustrated in Fig. 3.1. This structure employs symmetrically arranged co-directional half-wavelength open-loop resonators, with a total resonator length of approximately half-wavelength [23]. Through specific edge-coupling designs, including open-loop gaps and adjacent parallel coupling lines, a hybrid electric and magnetic coupling mechanism is successfully introduced. Electric coupling arises from electric field interactions at gap edges, while magnetic coupling is controlled by mutual inductance generated by current loops in the coupling lines. By precisely adjusting the gap dimensions S_1 and S_2, the ratio of these two coupling types can be dynamically balanced. The theoretical model for the coupling coefficient K_{12} is established using the odd-mode resonance frequency (f_{odd}), the even-mode resonance frequency (f_{even}), and the transmission zero frequency (f_m).

Simultaneously, this structure innovatively incorporates tunable external capacitors C_m at the input/output feedlines. Leveraging the impact of these capacitors on the imaginary part of the input admittance $Im[Y_{in}(x_0)]$, they directly control the external quality factor Q_{ext}. Combined with tuning the capacitance value C_l of varactor diodes within the resonant cavities, this enables precise bandwidth control. The unique no RF short circuited points design significantly enhances system stability, providing the physical foundation for implementing the hybrid coupling mechanism.

Subsequently, we propose higher-order coupled filter structures. These include filters based on parallel-coupled lines (J-inverters) and filters based on alternating J/K inverters. For the parallel-coupled line (J-inverter) approach, we designed and implemented second-order and third-order CABW filters with Butterworth responses, shown in Fig. 3.2a, b respectively [24]. These filters utilize symmetrically arranged varactor-loaded microstrip resonators, with input/output ports and inter-resonator coupling realized through J-inverters (physically implemented as parallel-coupled lines). The key design theory relies on precise analysis of the characteristics of

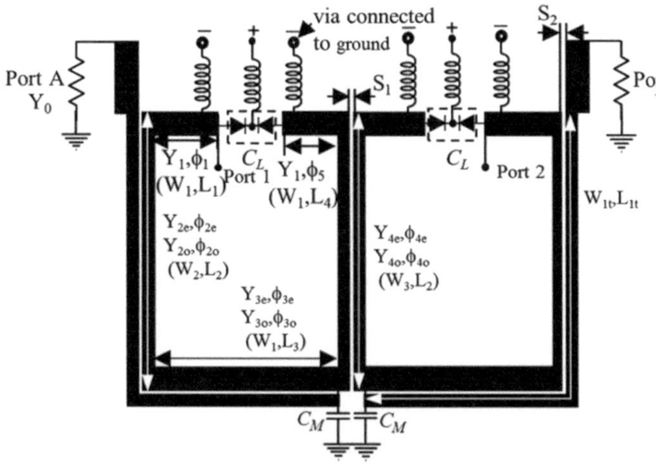

Fig. 3.1 Electrical circuit model of second-order electrically and magnetically mixed coupled microstrip open-loop tunable filter

varactor-loaded resonators. By extracting the physical gap dimension S of the coupling lines using full-wave electromagnetic simulation, the required, nearly constant normalized coupling coefficient values can be satisfied across the entire tuning range, ensuring stable absolute bandwidth during center frequency tuning.

For the more innovative alternating J/K inverter topology, this scheme is particularly suitable for compact layouts of even-order (e.g., fourth-order) filters. Its core innovation lies in directly realizing the K-inverter (providing negative magnetic coupling) using short-circuited transmission lines. As shown in Fig. 3.2c, the varactor-loaded resonators are reconfigured into compact hairpin forms. The crucial inter-stage coupling mechanism is provided by an integrated short-circuited stub

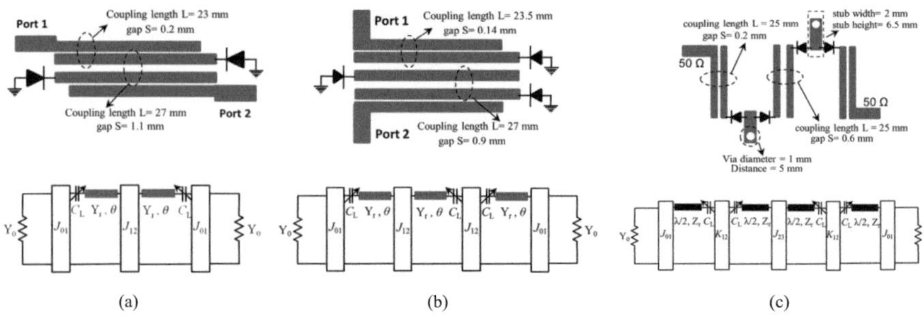

Fig. 3.2 Schematic and equivalent network with J inverters. **a** Second-order Butterworth tunable bandpass filter. **b** Third-order Butterworth tunable bandpass filter. **c** Fourth-order Butterworth tunable bandpass filter

connecting adjacent resonators and incorporating metalized vias. This stub essentially functions as a lumped K-inverter, replacing traditional distributed coupling structures. This integrated design merges coupling elements into the resonator units themselves, optimizing space utilization. A critical design consideration is that the K-inverter (the short-circuited stub), being part of the resonator, contributes to the effective electrical length determining the resonance frequency. The innovation of this topology lies in its integrated K-inverter based on a short-circuit structure, combined with traditional parallel-coupled-line J-inverters, handling input/output and partial inter-stage coupling, successfully establishing an effective synthesis path for higher-order filters.

Building on this, we further developed a compact sixth-order filter topology [25]. This topology combines J-inverters and K-inverters in a square layout with integrated cross-coupling paths, as shown in Fig. 3.3. The design selectively leverages the properties of anti-parallel and parallel coupled microstrip line structures. By precisely locating the coupling electrical length within specific operating regions, the susceptance value changes as the filter center frequency are tuned, achieving the goal of constant absolute bandwidth without needing separate components to control coupling strength. The core innovation involves reshaping the resonators into a compact square arrangement and ingeniously utilizing spatial relationships to introduce cross-coupling paths between Resonators 1 and 6, and between Resonators 2 and 5. These paths are equivalent to integrated K-negative inverters, working together with J-inverters, realized by parallel-coupled microstrip lines, forming the main coupling paths to create a hybrid J/K negative inverter network structure. This topology enhances out-of-band suppression by realizing multiple transmission zeros for a given filter order and maintains stable absolute bandwidth over a wide tuning range due to its structural advantages.

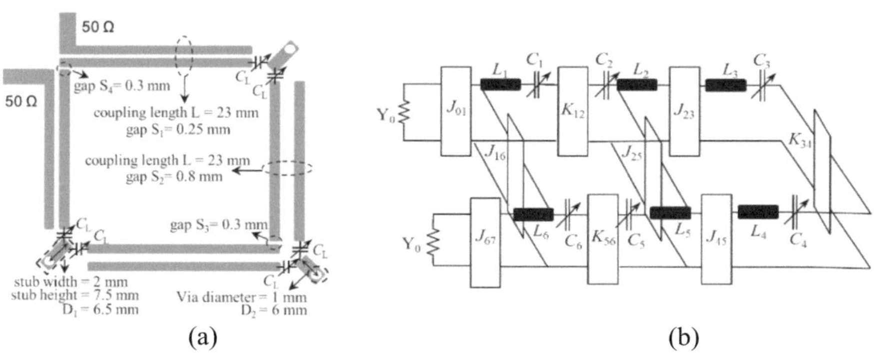

Fig. 3.3 Six-order constant absolute bandwidth tunable filter. **a** Schematic layout. **b** Alternative J/K inverters equivalent network

Addressing the urgent demand for constant fractional bandwidth (CFBW) in modern multi-band communication systems, we explored stabilizing relative bandwidth in higher-order structures [26, 27]. As shown in Fig. 3.4, by innovatively introducing cascaded three-cavity and four-cavity topologies into the realm of microstrip tunable filters, we achieved continuously tunable designs based on cross-coupling schemes. The key advantage of these topologies is their flexibility in constructing cross-coupling paths. Specifically, by loading tuning capacitors onto shorted-line resonators and incorporating magnetic coupling microstrip structures, the coupling paths not only support the main signal path but also introduce electric coupling between non-adjacent resonators. This dynamically generates transmission zeros at the passband edges, improving filter selectivity. More importantly, by optimizing the physical parameters of the magnetic coupling structure, the magnetic coupling coefficient K_m can be maintained highly stable during tuning. Concurrently, the design of the input/output port networks ensures a relatively constant external quality factor Q_{ext} across the tuning range. This filter structure overcomes the challenge of bandwidth drift in conventional tunable filters. Furthermore, this design simplifies the control strategy: center frequency tuning is achieved via the tuning capacitor C_t loaded at the resonator terminals, while the reconfigurability of transmission zeros relies on the independently tunable electric coupling coefficient controlled by the cross-coupling capacitor C_{tc}. Consequently, only two bias voltages are needed to manage both frequency position and edge selectivity, significantly reducing system complexity.

To further meet the unique application requirement of flexible, independent bandwidth tuning at a fixed center frequency, based on a deep understanding of resonant mode behavior, we propose two innovative multi-mode resonator topologies. The first filter is the triple-mode resonator topology [28]. As shown in Fig. 3.5, this structure skillfully utilizes a symmetrically arranged resonator unit whose core contains

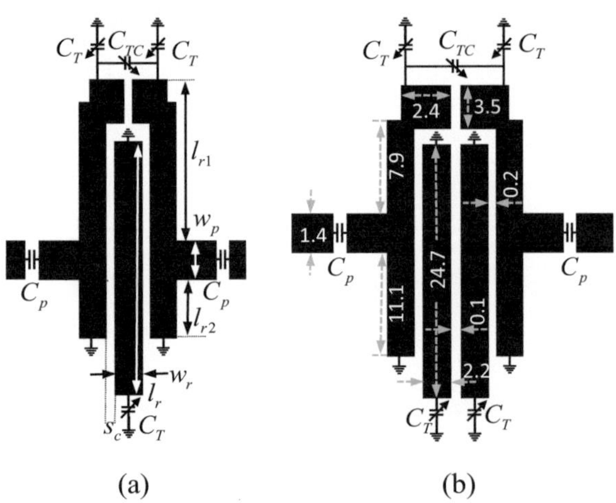

Fig. 3.4 Dimensions of the proposed tunable BPF. **a** Proposed 3-pole tunable BPF. **b** Proposed 4-pole tunable BPF

a main transmission line supporting an odd-mode resonance. This odd-mode resonance frequency is strictly constrained by its physical length and thus remains fixed, providing a stable reference point for independently tuning the upper and lower passband edges. At the symmetric center point, two open stubs loaded with varactor diodes enable separate control of two independent even-mode resonance frequencies. These two tunable even modes directly correspond to the low cutoff edge and high cutoff edge of the passband, and their movement defines the bandwidth tuning range. A key design aspect: the varactors on each tuning stub not only adjust their corresponding even-mode frequency but also cooperatively generate two associated transmission zeros (TZs). These TZs are strategically placed near the low and high sides of the passband, crucial for suppressing adjacent signals and enhancing stopband rejection. Notably, by changing the independent bias voltages applied to the two varactors, the lower-order even mode (and its associated low-side TZ) and the higher-order even mode (and its associated high-side TZ) can be moved separately. This enables completely independent and continuous control of the low and high cutoff frequencies of the passband while the center frequency remains fixed. This structure provides an intuitive and efficient mechanism, requiring only a few tuning elements (varactors), to offer flexible bandwidth adaptation at specific operating frequencies while effectively suppressing harmonics.

As shown in Fig. 3.6, the second filter is the quadruple-mode stub-loaded resonator topology [29]. Through more refined mode decoupling analysis, this topology reveals how varactors at different loading positions dominate distinct resonant modes and edge transmission zeros, achieving continuous bandwidth tuning at a fixed center

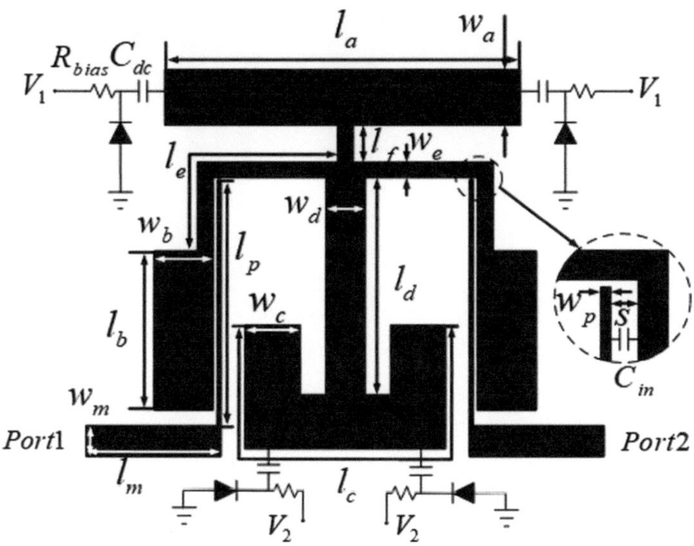

Fig. 3.5 Layout of the bandwidth tunable BPF with harmonic suppression

frequency. The structure uses two sets of independent varactor diodes loaded symmetrically at the terminals of open stubs to achieve decoupled control over four resonant modes (two odd modes and two even modes) and TZs at both passband edges. Research shows that varactor C_1 loaded on specific stubs primarily influences the odd-mode frequency and the upper-edge transmission zero (TZ_1), while varactor C_2 loaded on another symmetric stub mainly controls the even-mode frequency and the lower-edge transmission zero (TZ_2). This relationship allows designers to precisely shift the high-frequency edge and low-frequency edge of the passband separately by independently adjusting the bias voltages of C_1 and C_2, all while maintaining center frequency stability. For example, increasing the capacitance of C_1 lowers the odd-mode frequency and shifts TZ_1 downward, thereby reducing the upper cutoff frequency. Correspondingly, increasing the capacitance of C_2 lowers the even-mode frequency and shifts TZ_2 downward, reducing the lower cutoff frequency. Through this mechanism for independent, continuous adjustment of the two edge frequencies, the filter bandwidth can be continuously expanded or contracted without moving the center frequency.

These two multi-mode resonator structures open new technical pathways for applications requiring flexible bandwidth adaptation on a fixed operating channel.

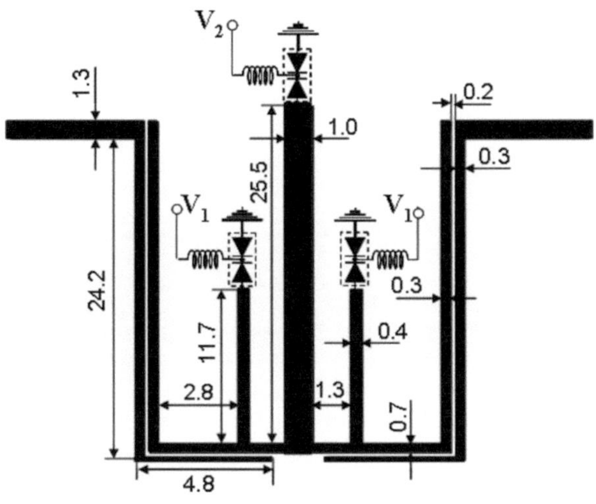

Fig. 3.6 Layout of the proposed tunable filter (all dimensions are in millimeters)

3.3 Theoretical Analysis

3.3.1 Tunable Filter Using Varactor-Loaded Open-Loop Resonators

The total length of the resonator in Fig. 3.1 is around a half-wavelength and the structure is symmetrical about the vertical axis. The coexisting electric coupling E_C and magnetic coupling M_C are sensitive to the sides with the open gaps and the adjacent coupling lines. The line with the width of W_{1t} and the length of L_{1t} couples to the line with the width of W_2 and constructs the input (output). The matching capacitor C_M is used to adjust the value of external quality Q_{ext}.

The input admittance of the uncoupled capacitively-loaded loop resonator seen form port A can be calculated by Y admittance using full-wave simulator. The resonant condition be found using

$$\mathrm{Im}[Y_{in}(\omega_0)] = 0 \tag{3.1}$$

The bandwidth characteristic of tunable filter is depended on the coupling coefficient k_{12} and Q_{ext}. For the second-order coupled resonator filter, the coupling coefficient k_{12} and external quality Q_{ext} can be determined as

$$k_{12} = \frac{FWB}{\sqrt{g_1 g_2}} \tag{3.2}$$

$$Q_{ext} = \frac{g_0 g_1}{FWB} \tag{3.3}$$

where FBW represent the fractional bandwidth of the passbands. g_0 and g_1 are the element values of the prototype lowpass filter. The coupling coefficient k of the proposed structure can be calculated by

$$k = \frac{M_c - E_c}{1 - M_c E_c} \tag{3.4}$$

where

$$M_c = \frac{1}{2}\left(\frac{\omega_{odd}^2 - \omega_0^2}{\omega_{odd}^2 - \omega_m^2} + \frac{\omega_{even}^2 - \omega_0^2}{\omega_m^2 - \omega_{even}^2}\right) \tag{3.5}$$

$$E_c = \frac{\omega_m^2}{2\omega_0^2}\left(\frac{\omega_0^2 - \omega_{odd}^2}{\omega_m^2 - \omega_{odd}^2} + \frac{\omega_0^2 - \omega_{even}^2}{\omega_{even}^2 - \omega_m^2}\right) \tag{3.6}$$

The ω_{odd}, ω_{even}, ω_m and ω_0 represent the odd mode, even mode resonant frequency, transmission zero frequency and self-resonant frequency, respectively. f_{odd}, f_{even} and f_m can be directly observed from the S-parameter response between two ports under

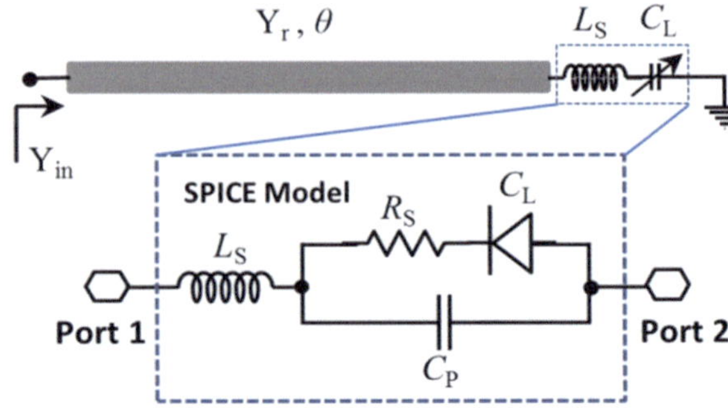

Fig. 3.7 Schematic of the presented varactor-loaded microstrip resonator

weakly coupling condition. The values of Q_{ext} can be numerically calculated under a given condition (e.g., known values of C_L and C_M). The full-wave simulation of the left half microstrip structure of Fig. 3.7 (without C_L and C_M) is performed using SONNET and the S-parameters are extracted. Then, the S-parameters are imported into Advanced Design System (ADS) and loaded with C_L and C_M. The value of external coupling Q_{ext} can be calculated as follows

$$\mathrm{Im}[Y_{11}(\omega_0)] = 0, \quad Q_{ext} = \frac{b}{\mathrm{Re}[Y_{11}(\omega_0)]} = \frac{g_1 g_2}{\Delta} \tag{3.7}$$

where

$$b = \mathrm{Im}\left[\frac{\omega_0}{2}\frac{\partial Y_{11}(\omega_0)}{\partial \omega}\right] \tag{3.8}$$

3.3.2 *Tunable Filter Using Varactor-Loaded Microstrip Resonator*

For the load resonator of a varactor diode shown in Fig. 3.7, the resistance R_s, capacitance C_P and inductance L_S in Fig. 3.7 represent the parasitic effects of a packaged varactor. Compared to the junction capacitance C_L, the capacitance C_P is negligibly small. The value of R_s is reasonably small and it may affect the insertion loss of a resultant circuit. Thus, the SPICE model of a varactor diode can be simplified as an inductance in series with a capacitance and the input impedance Y_{in} at the left-end, looking to the right-end, in Fig. 3.7 can be derived as

$$Y_{in} = Y_r \frac{j\frac{\omega C_L}{1-\omega^2 L_S C_L} + jY_r \tan\theta}{Y_r - \frac{\omega C_L}{1-\omega^2 L_S C_L}\tan\theta} \tag{3.9}$$

According to the resonant condition, $\mathrm{Im}[Y_{in}] = 0$ leads to

$$C_L = -\frac{Y_r \tan\theta_0}{\omega_0 - \omega_0^2 L_S Y_r \tan\theta_0} \tag{3.10}$$

The susceptance slope b_r of the varactor-loaded microstrip resonator is defined and derived as

$$\frac{b_r}{Y_r} = \frac{1}{Y_r}\frac{\omega_0}{2}\frac{dB(\omega)}{d\omega}\bigg|_{\omega=\omega_0} = \frac{\theta_0 - 1/2\sin 2\theta_0 + 2\omega_0 L_S Y_r \sin^2\theta_0}{2} \tag{3.11}$$

where θ_0 is the mid-band electrical length of the resonator.

Taking the second-order Butterworth adjustable filter shown in Fig. 3.2a as an example, in its J-type inverter equivalent network, the susceptance of J inverter can be calculated by taking the low-pass cutoff frequency $\Omega_c = 1$,

$$\frac{J_{01}}{\sqrt{Y_0 Y_r}} = \sqrt{\frac{BW}{f_0}} \cdot \sqrt{\frac{b_r}{Y_r}} \cdot \sqrt{\frac{1}{g_0 g_1}} \tag{3.12}$$

$$\frac{J_{12}}{Y_r} = \frac{BW}{f_0} \cdot \frac{b_r}{Y_r} \cdot \sqrt{\frac{1}{g_1 g_2}} \tag{3.13}$$

where BW stands for the absolute bandwidth, and g_n ($n = 0$, 1 and 2) are the three elements of the second-order low-pass prototype.

Substitute (3.11) to (3.12) and (3.13) to obtain

$$\frac{J_{01}}{\sqrt{Y_0 Y_r}} = \sqrt{\frac{BW}{f_0}} \cdot \sqrt{\frac{\theta_0 - 1/2\sin 2\theta_0 + 2\omega_0 L_S Y_r \sin^2\theta_0}{2}} \cdot \sqrt{\frac{1}{g_0 g_1}} \tag{3.14}$$

$$\frac{J_{12}}{Y_r} = \frac{BW}{f_0} \cdot \frac{\theta_0 - 1/2\sin 2\theta_0 + 2\omega_0 L_S Y_r \sin^2\theta_0}{2} \cdot \sqrt{\frac{1}{g_1 g_2}} \tag{3.15}$$

In theory, θ_0 can be calculated as

$$\theta_0 = \beta l = \frac{2\pi l\sqrt{\varepsilon_r} f_0}{\upsilon_0} \cdot \frac{180}{\pi} = A f_0 \tag{3.16}$$

where υ_0 and ε_r are the speed of the light in free space and permittivity of a dielectric substrate, f_0 is proportional to θ_0, and $A = \left(360L\sqrt{\varepsilon_r}\right)/v_0$ is constant value under certain physical length and dielectric permittivity.

To obtain the susceptance elements of the physical layout, J_{01} and J_{12} are then numerically extracted from the simulated two-port S-parameters based on the works in [30, 31]

$$\frac{J}{\sqrt{Y_1 Y_2}} = \frac{\sin\left(-\frac{\phi_1}{2}\right) + \overline{B}_{11}\cos\left(-\frac{\phi_1}{2}\right)}{\overline{B}_{12}\sin\left(-\frac{\phi_2}{2}\right)} \tag{3.17}$$

$$\phi_1 = M_1\pi + \tan^{-1}\left\{\frac{2\left(\overline{B}_{11} + \overline{B}_{22}\left|\overline{B}\right|\right)}{1 + \overline{B}_{22}^2 - \overline{B}_{11}^2 - \left|\overline{B}\right|^2}\right\} \tag{3.18}$$

$$\phi_2 = M_2\pi + \tan^{-1}\left\{\frac{2\left(\overline{B}_{22} + \overline{B}_{11}\left|\overline{B}\right|\right)}{1 + \overline{B}_{11}^2 - \overline{B}_{22}^2 - \left|\overline{B}\right|^2}\right\} \tag{3.19}$$

where $\overline{B}_{11} = B_{11}/Y_1$, $\overline{B}_{22} = B_{22}/Y_2$, $\overline{B} = \overline{B}_{11}\overline{B}_{22} - \overline{B}_{12}^2$, Y_1 and Y_2 are the characteristic admittance of two ports, M_1 and M_2 are the two non-negative integers.

The parameter *A* primarily depends on the effective electrical length of a microstrip line. The microstrip line is divided to two distinct parts with respect to the varactor diode, namely, open- and short-circuited parts. The short-circuited part not only comprises the through-via for varactor diode, but also serves as an inductive *K* inverter. Compared to the open-circuited part, the short-circuited part is much short in length. Thus, the short-circuited part could be considered as a lump inductor, and *A* fully depends on the length of the open-circuited part.

For the constant fractional bandwidth filter based on parallel coupled microstrip lines presented in Fig. 3.3, it is necessary to first analyze the working principle of parallel coupled microstrip lines. Figure 3.8a shows the physical structures of two types of parallel coupled microstrip lines: antiparallel and parallel coupled lines, the coupling lengths of the two structures are kept to be identical, $L_{C1} = L_{C2}$. According to Fig. 3.8a, b, the two J-susceptances vary as a quasi-periodical function of frequency. In particular, the value achieves the maximal peak at about 110° for anti-parallel-coupled line, whereas the coupling peak in pro-parallel-coupled line occurs at the frequencies of $\theta/2 = 45°$ and $\theta/2 = 135°$ while the coupling null is placed at the $\theta/2 = 90°$. In each period, the curves can be divided into three regions: region 1 (also C), region 2 (also A and D) and region 3 (also B and E), which represent the value of *J*-susceptance increases, remains stationary and decreases, respectively, as frequency increases.

According to filter synthesis theory [32], the values of *J*-susceptance is proportional to BWb_r/f_0, where BW, f_0, and b_r represent bandwidth, center frequency and susceptance slope, respectively. For a tunable resonator, the susceptance slope b_r is varied as center frequency. From the theory of resonators with loaded varactor diodes mentioned above, we can deduce the input admittances of the varactor-loaded $\lambda/4$ resonator and varactor-loaded $\lambda/2$ resonator, as Fig. 3.9 shown, seen form the input port, are

$$Y_{in1} = j\omega C_{L1} - jY_{r1}\cot\theta_1 \tag{3.20}$$

$$Y_{in2} = Y_{r2}\frac{j\omega C_{L2} + jY_{r2}\tan\theta_2}{Y_{r2} - \omega C_{L2}\tan\theta_2} \tag{3.21}$$

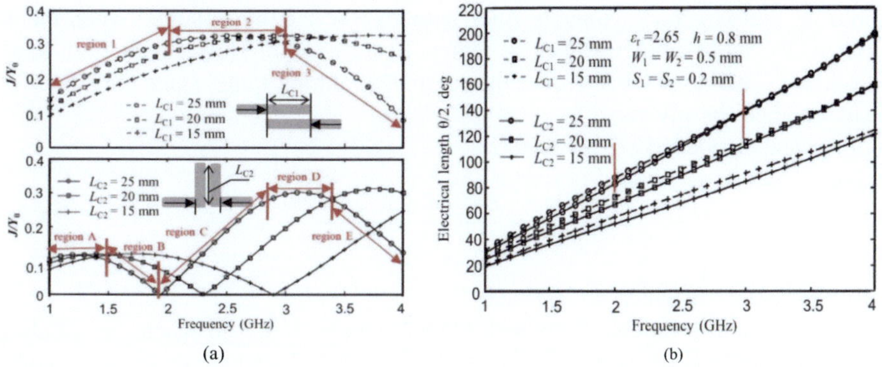

Fig. 3.8 Extracted J-inverter parameters of the two coupled microstrip lines. **a** Normalized J-susceptance. **b** Electrical length

where Y_{r1} and Y_{r2} are the characteristic admittances of the microstrip lines, θ_1 and θ_2 are the electric lengths, and C_{L1} and C_{L2} are the capacitance of varactors.

The susceptance slopes of the two varactor-loaded microstrip resonators are defined and derived as

$$\frac{b_{r1}}{Y_{r1}} = \frac{1}{Y_{r1}} \frac{\omega_0}{2} \frac{dB(\omega)}{d\omega}\bigg|_{\omega=\omega_0} = \frac{\cot\theta_0 + \theta_0 \csc^2\theta_0}{2} \tag{3.22}$$

$$\frac{b_{r2}}{Y_{r2}} = \frac{1}{Y_{r2}} \frac{\omega_0}{2} \frac{dB(\omega)}{d\omega}\bigg|_{\omega=\omega_0} = \frac{\theta_0 - 1/2\sin 2\theta_0}{2} \tag{3.23}$$

For varactor-loaded λ/4 resonator,

$$\frac{BWb_r}{f_0} = \frac{BW}{f_0} \frac{\cot\theta_0 + \theta_0 \csc^2\theta_0}{2} \tag{3.24}$$

where $\theta_0 = \beta l = 2\pi l \sqrt{\varepsilon_r} f_0 / \upsilon_0$ and the value is smaller than π/4.

It is observed that a tunable filter with CABW can be designed when coupling electrical lengths work at region B in Fig. 3.8a, which can be realized by properly

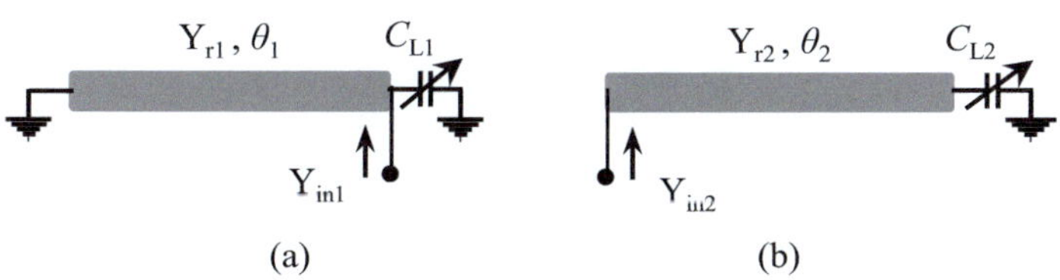

Fig. 3.9 Two typical tunable resonators. **a** Varactor-loaded λ/4 resonator. **b** Varactor-loaded λ/2 resonator

choosing the varactor; a tunable filter with CFBW can be obtained when the J-susceptance with smaller slopes in region B are chosen. However, it is difficult to achieve CABW or CFBW over the reasonable tuning range since the values of J-susceptance sharply vary.

For varactor-loaded λ/2 resonator,

$$\frac{BWb_r}{f_0} = \frac{BW}{f_0}\frac{\theta_0 - 1/2\sin 2\theta_0}{2} \tag{3.25}$$

where θ_0 locates between π/4 and π/2.

It is seen that a tunable filter with CABW can be achieved when the coupling electrical lengths worked at region 1 (also C) in Fig. 3.8a, which can be realized by properly choosing the varactor; a tunable filter with CFBW can be designed when the J-susceptance with bigger slopes in region A (also C) are chosen.

3.3.3 Tunable Filter Using Cascade Triplet and Quadruplet Topology

Figure 3.4a, b show the constant fractional bandwidth adjustable filters based on cross coupling. Now we take the 3-pole tunable filter as an example to analyze its working principle. The filter shown in Fig. 3.10 consists of three shorted resonators loaded with tunable capacitors C_T and in/out ports.

The coupling scheme and matrix of the proposed tunable filter are shown in Fig. 3.10. Coupling matrix M_{3pole} analyzed for this arrangement also considers the source/load couplings. Therefore, in an N order filter, the coupling matrix has N + 2 rows and N + 2 columns. A symmetrical coupling is obtained when a symmetrical

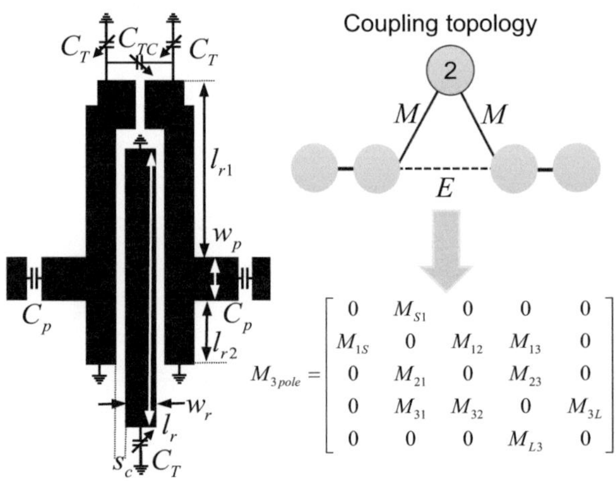

Fig. 3.10 Configuration of the proposed 3-pole tunable bandpass filter based on the cross-coupling cascade triplet (CT) topology and its coupling scheme, M matrix

layout is assumed in the structure, thus, $M_{AB} = M_{BA}$, $M_{12} = M_{23}$, and $M_{S1} = M_{L3}$. Furthermore, the main couplings between resonators 1 and 2, 2 and 3 are considered as magnetic coupling K_m, while the cross coupling between resonators 1 and 3 is electric coupling K_e. Due to this cross-coupling design, more TZs are generated around the passband to improve the selectivity of the filter.

The coupling matrix coefficients are also frequency independent and the theoretical frequency responses of the filter can be obtained by using the following equations:

$$\left[-jR + \omega' W + M\right][I] = [A][I] \tag{3.26}$$

where R is a (N + 2, N + 2) matrix, in which only nonzero entries are $R_{11} = R_i$, and $R_{N+2,N+2} = Ro$, both normalized to 1 with 50 Ω reference, W is a (N + 2, N + 2) identity matrix with $W_{11} = W_{N+2,N+2} = 0$, ω' is the low-pass prototype normalized angular frequency given by

$$\omega' = \frac{\omega_c}{\Delta\omega}\left(\frac{\omega}{\omega_c} - \frac{\omega_c}{\omega}\right) \tag{3.27}$$

where ω_c is the center frequency of the filter, $\Delta\omega$ is the bandwidth, and M is the coupling matrix. Finally, the S-parameters can be deduced by

$$S_{21} = -2j\left[A^{-1}\right]_{n+2,1} \quad S_{11} = 1 + 2j\left[A^{-1}\right]_{11} \tag{3.28}$$

Derived from the extracted method of the coupling coefficients and Q_{ext} presented in [32] and Eqs. (3.26), (3.27), (3.28), an initial coupling matrix can be established. Then, its coefficients can be adjusted so that the frequency response can match the desired specifications and guide the filter design in the EM simulation. Referring to the CT topology presented in Fig. 3.10, a filter with three TZs around the passband centered at 1.45 GHz with a 12.5% 3-dB fractional bandwidth is modeled by the following coupling matrix.

$$M_{3pole} = \begin{bmatrix} 0 & 0.908 & 0 & 0 & 0 \\ 0.908 & 0 & 0.708 & -0.350 & 0 \\ 0 & 0.708 & 0 & 0.708 & 0 \\ 0 & -0.350 & 0.708 & 0 & 0.908 \\ 0 & 0 & 0 & 0.908 & 0 \end{bmatrix} \tag{3.29}$$

Considering the microstrip realization, some coupling coefficients have been already corrected in (3.29). Based on M_{3pole}, we can achieve the frequency responses of the proposed 3-pole filter. Moreover, by changing the relevant coefficients, the theoretical frequency responses can be analyzed as follows.

A. ***Frequency Responses Against*** $\boldsymbol{K_m}$

The variation of K_m, represented by x in (3.30), indicates the possible responses where the bandwidth of filter can be tuned without changing the center frequency. As desired, three TZs are located next to the passband to improve the selectivity of the filter. Furthermore, it can be seen that the overall frequency responses remain stable while changing the bandwidth.

$$M_{3pole} = \begin{bmatrix} 0 & 0.908 & 0 & 0 & 0 \\ 0.908 & 0 & 0.708+x & -0.350 & 0 \\ 0 & 0.708+x & 0 & 0.708+x & 0 \\ 0 & -0.350 & 0.708+x & 0 & 0.908 \\ 0 & 0 & 0 & 0.908 & 0 \end{bmatrix} \quad (3.30)$$

B. ***Frequency Responses Against*** $\boldsymbol{K_e}$

Meanwhile, the variation of K_e, represented by y in (3.31), conveys that the transmission zeros can be tuned while keeping the bandwidth and the center frequency almost unchanged. Obviously, the locations of TZs in the low-side of the passbands can be tuned by varying M_{13}. However, both of passband and the TZ in the up-side of the passband show a small variation, which can be taken as constant. To include, the selectivity of low-side of the passband can be freely tuned by varying M_{13} with introducing little influence on other parameters of the passband. This unique property creates an additional design freedom and application scenario.

$$M_{3pole} = \begin{bmatrix} 0 & 0.908 & 0 & 0 & 0 \\ 0.908 & 0 & 0.708 & -0.350+y & 0 \\ 0 & 0.708 & 0 & 0.708 & 0 \\ 0 & -0.350+y & 0.708 & 0 & 0.908 \\ 0 & 0 & 0 & 0.908 & 0 \end{bmatrix} \quad (3.31)$$

C. ***Frequency Responses Against*** $\boldsymbol{M_{S1}}$

The variation of M_{S1}, represented by z in (3.32), addresses the effect of Q_{ext} on the frequency responses. The shape of the filter remains similar when M_{S1} varies within a certain range. When Q_{ext} is higher, the selectivity is higher; the passband becomes flatter while Q_{ext} is lower.

$$M_{3pole} = \begin{bmatrix} 0 & 0.908+z & 0 & 0 & 0 \\ 0.908+z & 0 & 0.708 & -0.350 & 0 \\ 0 & 0.708 & 0 & 0.708 & 0 \\ 0 & -0.350 & 0.708 & 0 & 0.908+z \\ 0 & 0 & 0 & 0.908+z & 0 \end{bmatrix} \quad (3.32)$$

The theoretical analysis made above shows some detailed effects of elements in M_{3pole} on the frequency response, which allows for achieving the desired filter response by optimizing those element values of M_{3pole}.

Unlike a fixed filter design, related parameters must satisfy the requirements to achieve desired properties during the tuning process. The following two equations must be satisfied for the proposed cross-coupling tunable filter design with constant bandwidth:

$$K_m = K_{12} = M_{12} \cdot FBW \quad Q_e = \frac{1}{FBW} \cdot \frac{1}{M_{S1}^2} \tag{3.33}$$

The initial main coupling coefficient K_m between resonator 1 and 2, 2 and 3 can be determined by the coupling matrix in (3.29). It can be inferred that to get a constant FBW in the tuning range, the coupling coefficients K_m should keep constant. Meanwhile, Q_e is inversely proportional to the FBW. Consequently, Q_e should be constant during the tuning range to maintain a CFBW tunable filter.

Figure 3.11 shows the equivalent circuit of the main coupling scheme between resonators 1 and 2, 2 and 3. The overall admittance matrix of the capacitively loaded coupled resonators is

$$Y = \begin{bmatrix} \frac{Y_{ine}+Y_{ino}}{2} & \frac{Y_{ine}-Y_{ino}}{2} \\ \frac{Y_{ine}-Y_{ino}}{2} & \frac{Y_{ine}+Y_{ino}}{2} \end{bmatrix} \tag{3.34}$$

where $Y_{ine,o} = j\omega C_T + Y_{re,o}$, $Y_{re,o} = jY_S \frac{Y_S \tan\theta_S - Y_{Ce,o}\cot\theta_{Ce,o}}{Y_S + Y_{Ce,o}\cot\theta_{Ce,o}\tan\theta_S}$.

For the above Y-network, two conditions must be satisfied, namely resonance condition and coupling condition. It can be formulated as

$$\text{Im}[Y_{11}] = 0, \quad \frac{\text{Im}[Y_{12}]}{b} = K_m \tag{3.35}$$

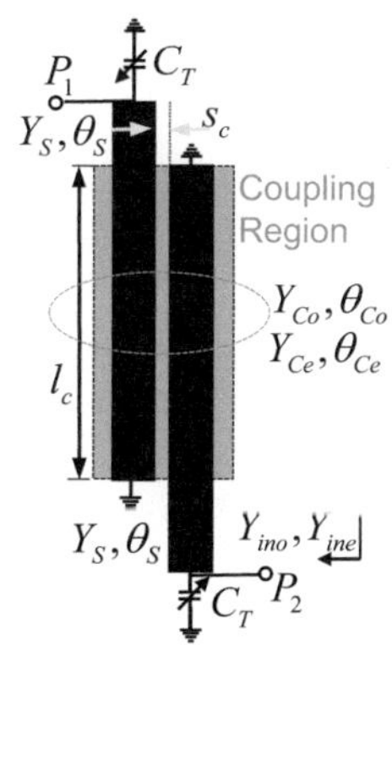

Fig. 3.11 Detailed structure of magnetic coupling part

where $b = \frac{\omega_0}{2} \frac{\partial \mathrm{Im}[Y_{11}|_{\omega=\omega_0}]}{\partial \omega}$.

Then, b can be transformed to the following type by applying Im[Y_{11}] = 0 to (3.35)

$$b = \mathrm{Im}\left[\frac{\omega_0}{4} \frac{\partial\left((Y_{re} + Y_{ro})|_{\omega=\omega_0}\right)}{\partial \omega} - \frac{(Y_{re} + Y_{ro})|_{\omega=\omega_0}}{4}\right] \tag{3.36}$$

Finally, we can get the equation for K_m as

$$K_m = \frac{\mathrm{Im}\left[\frac{Y_{re}-Y_{ro}}{2}|_{\omega=\omega_0}\right]}{\mathrm{Im}\left[\frac{\omega_0}{4} \frac{\partial\left((Y_{re}+Y_{ro})|_{\omega=\omega_0}\right)}{\partial \omega} - \frac{(Y_{re}+Y_{ro})|_{\omega=\omega_0}}{4}\right]} \tag{3.37}$$

According to (3.37), the requirement of K_m presented in (3.33) can be satisfied by properly choosing Y_s, θ_S, Y_{Co}, Y_{Ce}, θ_{Co} and θ_{Ce}. In brief, K_m can be realized by choosing the even–odd-mode coupled section l_c. Since the electrical length of these resonators is smaller than 90°, the net coupling is magnetic as desired before.

Figure 3.12 shows the equivalent LC circuit of the cross-coupling scheme. The tunable capacitor network can be transformed into a simple network based on odd mode equivalent capacitor C_{odd} and even mode equivalent capacitor C_{even}, K_e can be written as

$$K_e = \frac{C_{\mathrm{odd}} - C_{\mathrm{even}}}{C_{\mathrm{odd}} + C_{\mathrm{even}}} = \frac{C_{\mathrm{TC}}}{\frac{C_{\mathrm{T}}C_{\mathrm{R}}}{C_{\mathrm{T}}+C_{\mathrm{R}}} + C_{\mathrm{TC}}} \tag{3.38}$$

where $C_{\mathrm{even}} = \frac{C_{\mathrm{T}}C_{\mathrm{R}}}{C_{\mathrm{T}}+C_{\mathrm{R}}}$, $C_{\mathrm{odd}} = \frac{C_{\mathrm{T}}C_{\mathrm{R}}}{C_{\mathrm{T}}+C_{\mathrm{R}}} + 2C_{\mathrm{TC}}$.

C_R and L_R are the equivalent capacitance and inductance of the shorted microstrip line (Y_T, θ_T), respectively. C_T and C_{TC} are the capacitance of the tunable varactor diodes. K_e is influenced by C_T and C_{TC}. To be more specific, Fig. 3.13 displays the electric coupling coefficient K_e versus C_T. The absolute value of K_e increases while C_{TC} increases. In addition, the variation range of K_e will be enlarged with the

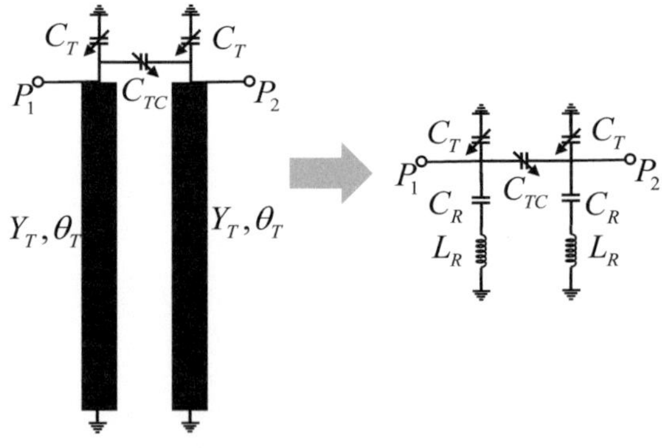

Fig. 3.12 Detailed structure and its equivalent LC circuit of electric coupling part

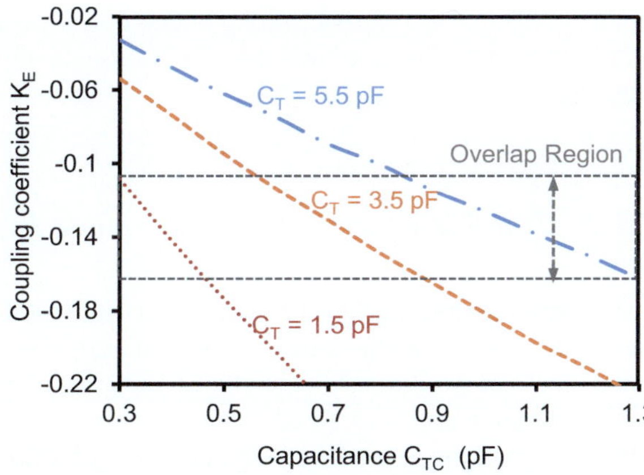

Fig. 3.13 Electric coupling coefficient K_e versus different coupling section C_{TC} and tunable capacitance C_T

decrease of C_T. In order to achieve the same tunability of TZs while tuning the center frequency of the filter, each variation range of K_e at different value of C_T should contain the required value range, which is indicated by the overlap region shown in Fig. 3.13.

Adopting this method, the TZs can be tuned to have the same roll-off response at each point while tuning the center frequency of the filter. Consequently, by properly choosing the ranges of C_{TC}, C_T and C_R, the requirements of K_e for a tuning design can be satisfied.

D. ***External Coupling of the Filter***

Figure 3.14 shows the input/output network of the proposed tunable filter. Then, the external quality Q_e can be written as a function of admittance Y_{in}

$$Q_e = \frac{\omega_0}{2Y_T}\frac{\partial \mathrm{Im}[Y_{in}(\omega_0)]}{\partial \omega} \tag{3.39}$$

where $Y_{in} = \frac{j\omega C_p Y_{load}}{j\omega C_p + Y_{load}}$, $Y_{load} = jY_T\left(\frac{\omega C_T + Y_T \tan\theta_1}{Y_T - \omega C_T \tan\theta_1} - \cot\theta_2\right)$.

It can be found that Q_e is affected by capacitor C_p, electrical lengths θ_1 and θ_2. Then, Q_e can be designed to meet the requirements of CFBW in (3.33) by properly optimizing those three parameters.

Finally, with all these requirements satisfied, a tunable 3-pole bandpass filter with CFBW based on C_T topology can be implemented. The selectivity of the lower side of passband can also be tuned. Furthermore, only two bias voltages are needed to

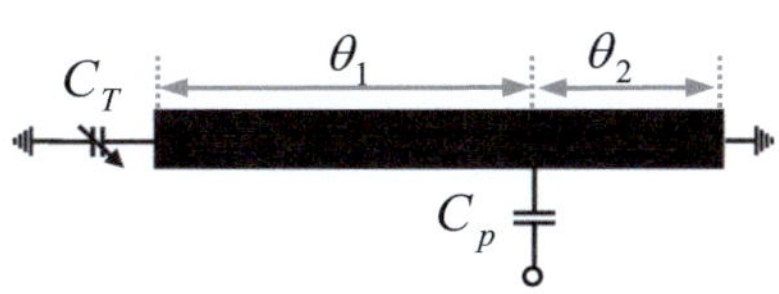

Fig. 3.14 Detailed structure of input network

complete the tuning process, among which one is used to tune the center frequency and the other is utilized to tune the TZs. Consequently, the tunability complexity is reduced.

3.3.4 Tunable Filter Using Multi-mode Resonator

For the bandwidth-tunable filter based on a triple-mode resonator shown in Fig. 3.5, due to its overall structure being symmetrically distributed relative to the central plane, this type of filter can be analyzed using the even–odd mode analysis method. As depicted in Fig. 3.15, C_1 and C_2 are the variable loading capacitance of the stubs. All the width of the resonators is chosen as w so that all the characteristic admittances is Y_1, and l_a, l_b, l_c, l_d, l_e refer to the length parameter of the resonator in Fig. 3.5. The resonant conditions can be achieved and expressed as

$$Y_{ino} = -jY_1 \cot \phi_1 \tag{3.40}$$

$$Y_{ine} = Y_1 \frac{(Y_{L11} + Y_{L12}) + jY_1 \tan \phi_1}{Y_1 + j(Y_{L11} + Y_{L12}) \tan \phi_1} \tag{3.41}$$

where $Y_{L11} = \frac{Y_1}{2} \cdot \frac{Y_{L1} + (jY_1 \tan \phi_2)/2}{Y_1/2 + jY_{L1} \tan \phi_2}$, $Y_{L12} = \frac{Y_1}{2} \cdot \frac{Y_1 \frac{Y_{L2} + jY_1 \tan \phi_{52}}{Y_1 + jY_{L2} \tan \phi_{52}} + (jY_1 \tan \phi_4)/2}{Y_1/2 + jY_1 \frac{Y_{L2} + jY_1 \tan \phi_{52}}{Y_1 + jY_{L2} \tan \phi_{52}} \tan \phi_4}$, $Y_{L1} = jY_1 \frac{\omega C_1 + Y_3 \tan(\phi_3/2)}{Y_1 - \omega C_1 \tan(\phi_3/2)}$, $Y_{L2} = j(\omega C_2 + Y_1 \tan \phi_{51})$.

The transmission zero frequencies are obtained when $Y_{21} = Y_{12} = 0$. The filter TZs can be calculated by

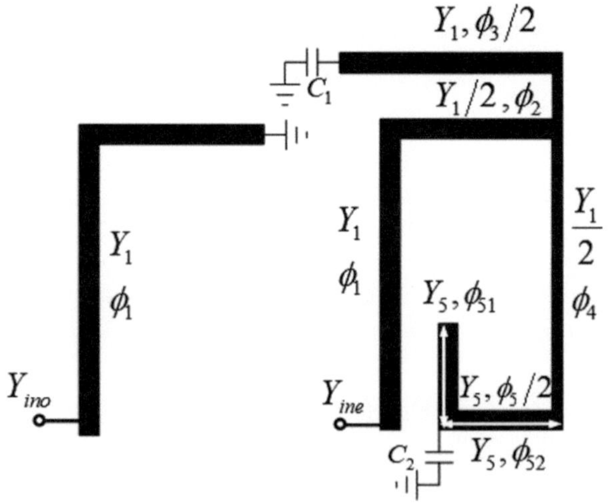

Fig. 3.15 Odd and even mode equivalent circuit of the proposed triple-mode resonator

$$Y_{12} = \frac{Y_1^2}{Y_{L11} \cdot \sin^2 \phi + Y_{L12} \cdot \sin^2 \phi - jY_1 \sin 2\phi} \tag{3.42}$$

where $Y_{L11} \to \infty$, $Y_{L12} \to \infty$.

It can be observed from (3.40) and (3.41) that the stepped-impedance varactor-loaded stub at the center plane merely controls resonant frequencies of even excitation. In order to achieve the detailed scheme of the resonant modes and TZs, an instructive method is adopted [13], as shown in Fig. 3.16. In this method the resonator is driven through a weak coupling. In Fig. 3.16a, the common characteristic can be obtained that the even-mode resonant frequencies decrease while increasing the length l_a or l_d. The conclusion can be made here that the lower resonant frequency is the first even-mode frequency which is controlled by inner stub. The middle resonant frequency is the first odd-mode frequency which is controlled by the l_b shown by (3.40). Then the third resonant frequency is the second even-mode frequency, which is controlled by the outer stub. By observing the simulated results presented in Fig. 3.16b, c, the TZ_1 is excited by the even-mode 1 and the odd-mode part, meanwhile, the TZ_2 is generated by the even-mode 2 and the odd-mode part. Finally, we can get the unique property of the tri-mode resonator that the even mode resonant frequencies and TZs can be controlled by adjusting the flexible parameters of the stubs, whereas the odd mode resonant frequency can hardly be changed.

The resonant frequencies and TZs can be calculated by the (3.40)–(3.42), as shown in Fig. 3.17. It is obvious that there are one odd mode, two even modes and two transmission zeros in the frequency range 1–3.5 GHz. In Fig. 3.17a, the higher even mode and TZ_2 move toward the lower frequency when $C_2 = 0.5$ pF and C_1 varies from 0.5 to 1.2 pF. Lower even mode and TZ_1 show slim variation and the odd mode remains constant. When $C_1 = 0.5$ pF and C_2 varies from 0.5 to 1.2 pF, as presented in Fig. 3.17b, the lower even mode and TZ_1 move toward the lower frequency, while the odd mode, higher even mode and TZ_2 remaining unchanged. Through the analysis above, the higher even mode and TZ_2 can be controlled by varying C_1. Simultaneous, the lower even mode and TZ_1 can be adjusted by varying C_2.

To suppress the second harmonic, through the analysis of the filter structure, the odd mode part offers the most important influence on the performance of the harmonic suppression. SIR (Stepped Impedance Resonator) is introduced to this filter to suppress the harmonic, as shown in Fig. 3.18a. Figure 3.18b shows the simulated frequency response of the filter against different W_m/W_n and a constant L_m/L_n. The second harmonic of the filter can be suppressed efficiently by tuning W_m while keeping $W_n = 0.8$ mm, $L_n = 8.2$ mm, $L_m = 11.2$ mm. The suppression effect on harmonic is very obvious while W_m increasing from 2 to 3 mm.

Similarly, the quadruple-mode resonator illustrated in Fig. 3.6—composed of a half-wavelength stepped impedance resonator (SIR), an open-circuit stub on the central plane, and two symmetric open-circuit stubs on both sides—also exhibits central-plane symmetry and can be analyzed using the odd–even mode method. By placing a short or open circuit at the symmetric plane of the resonator in Fig. 3.6, we obtain the circuit model for odd or even modes without the I/O coupling, as shown in

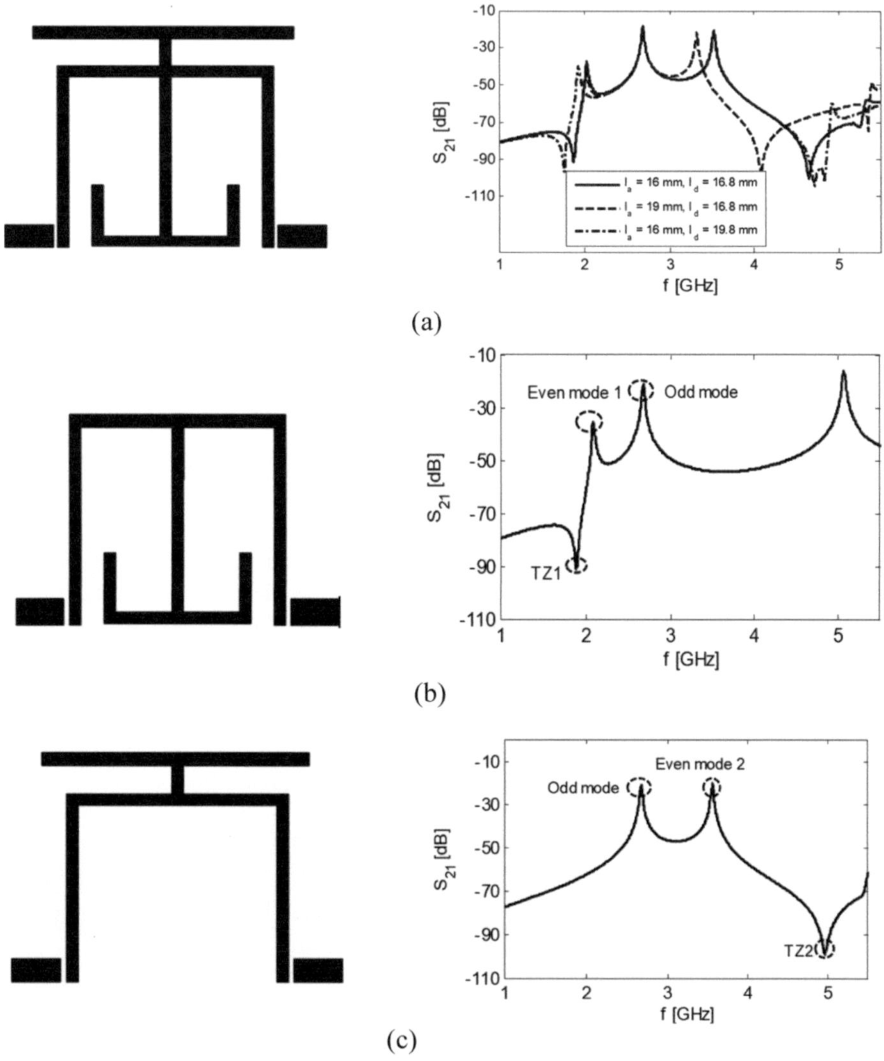

Fig. 3.16 Resonant-mode frequencies and TZs analysis. **a** Tri-mode resonator for different values of l_a or l_b. **b** Dual-mode resonator with inner stub. **c** Dual-mode resonator with outer stub

Fig. 3.19a, b. $Y_1, Y_2, Y_3, Y_4, \phi_1, f_{21}, f_{22}, f_3$ and f_4 refer to characteristic admittances and the electrical lengths for the transmission line sections. The resonant conditions can be achieved and expressed by

$$Y_{ino} = Y_1 \frac{Y_2(M + jY_2 \tan\phi_{21}) + jY_1(Y_2 + jM \tan\phi_{21}) \tan\phi_1}{Y_1(Y_2 + jM \tan\phi_{21}) + jY_2(M + jY_2 \tan\phi_{21}) \tan\phi_1} = 0 \qquad (3.43)$$

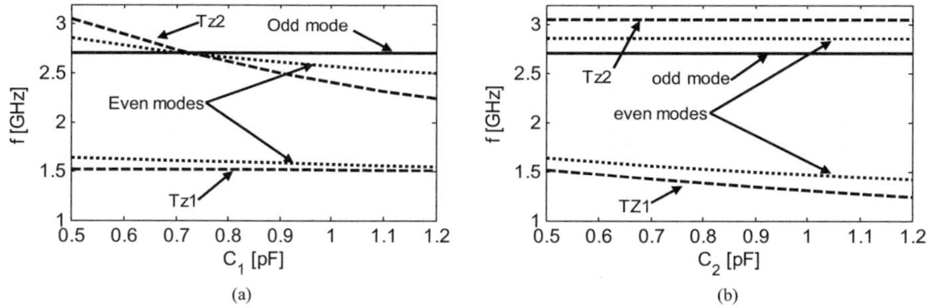

Fig. 3.17 Resonant-mode frequencies and TZs. **a** $C_2 = 0.5$ pF, varying C_1. **b** $C_1 = 0.5$ pF, varying C_2

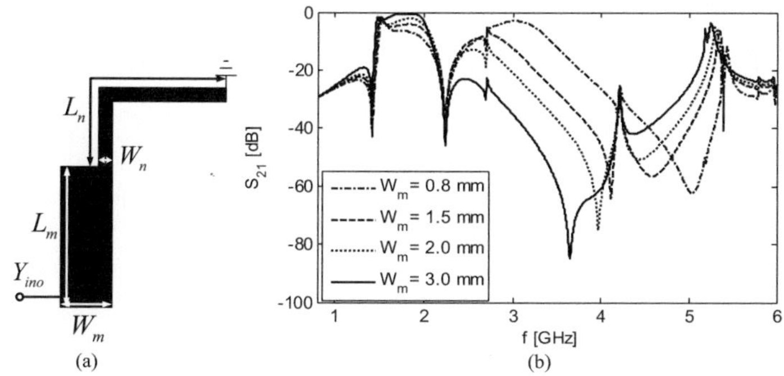

Fig. 3.18 Harmonic suppression analysis of the changed odd mode part. **a** Odd-mode part introduced with SIR. **b** Simulated results while fixed W_n, L_n and L_m, varying W_m

for odd modes, where $M = Y_{load1} - jY_2 \cot\phi_{22}$, $Y_{load1} = jY_3 \frac{\omega C_1 + Y_3 \tan\phi_3}{Y_3 - \omega C_1 \tan\phi_3}$ and

$$Y_{ine} = Y_1 \frac{Y_2\left(M' + jY_2 \tan\phi_{21}\right) + jY_1\left(Y_2 + jM' \tan\phi_{21}\right)\tan\phi_1}{Y_1(Y_2 + jM' \tan\phi_{21}) + jY_2(M' + jY_2 \tan\phi_{21})\tan\phi_1} = 0 \tag{3.44}$$

for even modes, where $M' = Y_{load1} + Y_2 \frac{Y_{load2} + jY_2 \tan\phi_{22}}{Y_2 + jY_{load2} \tan\phi_{22}}$ and $Y_{load2} = jY_4 \frac{\omega C_2 + Y_4 \tan\phi_4}{Y_4 - \omega C_2 \tan\phi_4}$.

The TZ frequencies are obtained when $Y_{21} = Y_{12} = 0$. Assuming $Y_1 = Y_2 = Y_0$ and $\theta_1 + \theta_{21} = \theta_{22} = \theta$, the filter TZs can be calculated by

$$\begin{aligned} Y_{12} = -\frac{Y_0^3}{j * Y_0^2 \sin 4\phi - 1/8 Y_{load1} Y_0 (1 + 4\cos 2\phi - 5\cos 4\phi)} &= 0 \qquad (3.45) \\ - 1/8 Y_{load2} Y_0 (2 + 2\sin 2\phi - 2\cos 4\phi + \sin 4\phi) \\ - j * 1/2 Y_{load1} \sin^2\phi \sin 2\phi (Y_{load1} + Y_{load2}) \end{aligned}$$

where

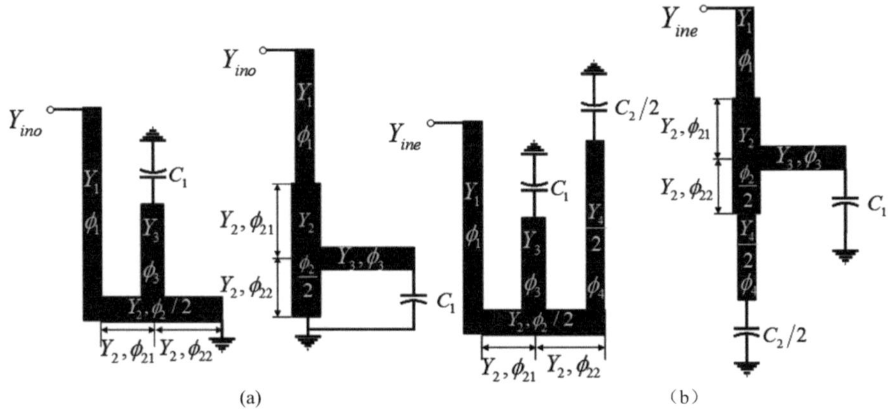

Fig. 3.19 Equivalent circuit of the quadruple-mode resonator. **a** Odd-mode. **b** Even-mode

$$Y_{load1} \to \infty, \ Y_{load2} \to \infty \tag{3.46}$$

From (3.43), (3.44), (3.46), resonant-mode frequencies and TZs versus the varactor C_1 and C_2 are shown in Fig. 3.20a, b, respectively. It is clear that there are two odd-modes, two even-modes and two TZs in the range of 1–3 GHz. As observed in Fig. 3.20a, when fixed $C_2 = 1.1$ pF and C_1 varies from 0.3 to 1.3 pF, the two odd-modes, two even-modes and TZ_1 move towards the lower frequency, whereas the TZ_2 remains stationary, and the higher odd-mode change larger than the other resonant modes. When fixed $C_1 = 0.5$ pF and C_2 varies from 0.3 to 1.3 pF as shown in Fig. 3.20b, the even-mode frequencies and TZ_2 are move towards the lower frequency, whereas the odd-mode frequencies and TZ_1 remain the same. Therefore, the higher odd-mode frequency and TZ_1 can be adjusted by varactor C_1 and the lower even-mode frequency and TZ_2 can be adjusted by varactor C_2. This unique property of the proposed quadruple resonator can be explored for designing filter with tunable bandwidth.

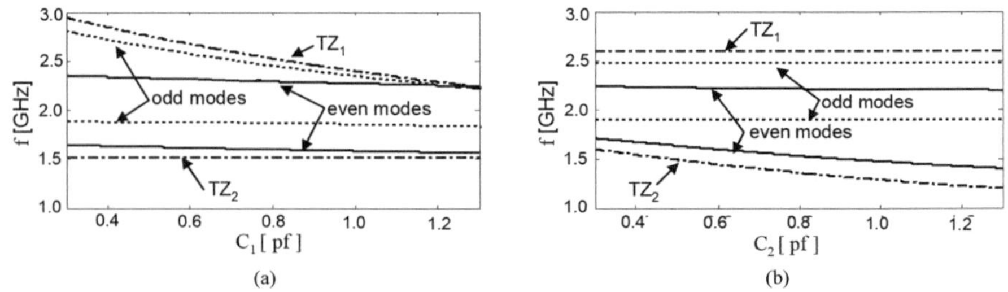

Fig. 3.20 Resonant-mode frequencies and TZs. **a** Fixed $C_2 = 1.1$ pF, varying C_1. **b** Fixed $C_1 = 0.5$ pF, varying C_2

3.4 Validation of the Proposed Tunable Bandpass Filters

3.4.1 *Validation of Constant Absolute Bandwidth (CABW) Tunable Filter*

Second-Order CABW Tunable Filter Based on Open-Loop Resonators

To validate the constant absolute bandwidth (CABW) design theory, we fabricated a second-order filter based on a co-directional open-loop resonator with hybrid coupling. As shown in Fig. 3.21a, this filter was implemented on a 0.508-mm-thick Rogers 5880 substrate ($\varepsilon_r = 2.55$) using copper etching technology. Two DC-isolated bias voltages were achieved by connecting SKYWORKS SMV-1405 varactor diodes in a common-cathode configuration. The single varactor capacitance measured 0.63 pF at 30 V reverse bias and 2.67 pF at 0 V bias. Three 82 nH RF chokes were implemented for DC biasing, effectively suppressing RF signal leakage in the bias network. A fixed 2.2 pF capacitor (C_m) was connected in parallel to the input feedline to regulate the external quality factor (Q_e) for maintaining constant absolute bandwidth. The electrical-to-magnetic coupling ratio satisfying the ABW condition was optimized by adjusting coupling-edge dimensions.

The measured 3-dB bandwidth and insertion loss are plotted in Fig. 3.21b. The center frequency was tuned from 607 to 942 MHz with insertion loss ranging from 2.5 to 3.3 dB. Notably, an absolute bandwidth of approximately 63 ± 7 MHz was maintained across the 642–942 MHz tuning range, demonstrating constant ABW performance. Insertion loss variation remained minimal throughout the entire tuning bandwidth.

We further designed and fabricated a second-order Butterworth tunable bandpass filter. As illustrated in Fig. 3.22a, this filter utilizes two symmetrically arranged varactor-loaded microstrip resonators with input/output (I/O) feedlines coupled through centrosymmetric parallel coupled lines functioning as J-inverters. The target

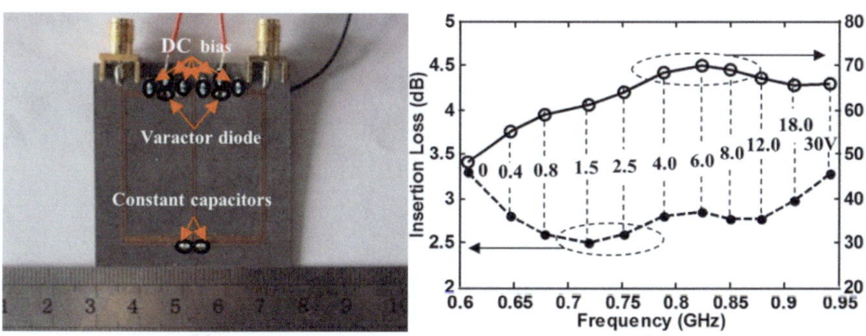

(a) (b)

Fig. 3.21 Photograph and measured results of the fabricated filter. **a** Photograph. **b** Measured insertion loss and 3-dB bandwidth of the tunable filter

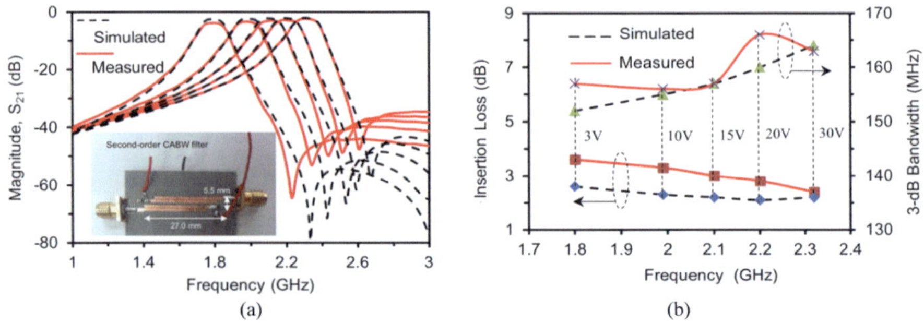

Fig. 3.22 Simulated and measured results. **a** S_{21}. **b** Insertion loss and 3-dB bandwidth of second-order tunable filter for bias voltage of: 3, 10, 15, 20, and 30 V (from left to right)

tuning range spans 1.80–2.32 GHz with a constant absolute bandwidth of 160 MHz. The filter was fabricated on an F4B-2 substrate ($\varepsilon_r = 2.65$, thickness = 0.5 mm), featuring resonators with 1.0 mm line width (characteristic impedance ≈ 60.54 Ω) and 27.0 mm length. J-inverter gap dimensions were numerically optimized to achieve target admittance values, resulting in an I/O coupling gap 0.2 mm and inter-resonator gap 1.1 mm. BB179 varactor diodes (capacitance range ≈ 2–20 pF) were implemented for tuning. Measured results in Fig. 3.22b demonstrate continuous center frequency tuning from 1.80 to 2.32 GHz while maintaining a 3-dB bandwidth of 160 ± 6 MHz. Insertion loss varied between 2.4 and 3.8 dB within the passband. The compact filter occupies 27.0 mm × 5.5 mm.

Higher-Order CABW Tunable Filter

Building upon the second-order J/K inverter topology for higher-order designs, we fabricated the third-order Butterworth tunable bandpass filter shown in Fig. 3.2b. This filter employs an identical J-inverter architecture targeting a 200 MHz constant bandwidth. It utilizes the same substrate material and resonator line width (characteristic impedance ≈ 60.54 Ω) as the second-order filter. The required J-inverter values were calculated from third-order low-pass prototype parameters and physically implemented through coupled-line structures. As demonstrated in Fig. 3.23a, b, measured results indicate:

Center frequency tuning: 1.76–2.26 GHz (5–30 V bias)
3-dB bandwidth: 192–202 MHz (maintained within target range)
Passband insertion loss: 3.8–4.8 dB
Return loss: 6.0–12.0 dB.

To minimize circuit area, we designed a fourth-order tunable bandpass filter using compact hairpin-shaped varactor-loaded microstrip resonators with alternating J/K inverter equivalent networks, targeting a 440 MHz bandwidth. For implementing complex coupling structures within limited space, K-inverters (enabling inductive coupling) were introduced between adjacent resonators through grounded stubs with metallized vias, while J-inverters retained for I/O and resonator coupling. Resonator

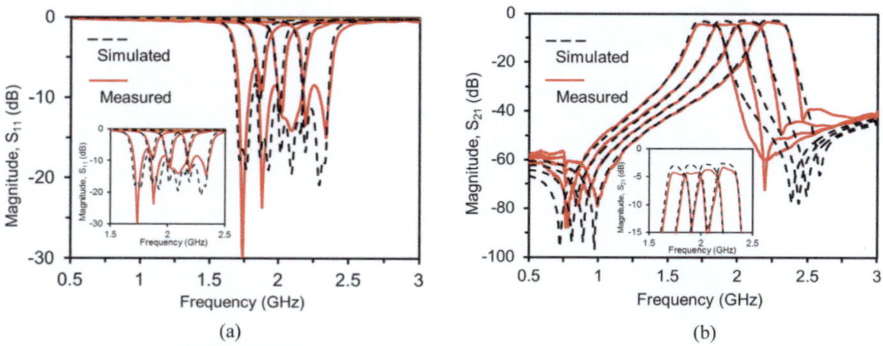

Fig. 3.23 Simulated and measured results. **a** S_{11}. **b** S_{21}. The bias voltage is 5, 10, 16, and 30 V (from left to right)

line width was reduced to 0.5 mm (characteristic impedance ≈ 87.54 Ω). As shown in Fig. 3.24a–c, measured results demonstrate continuous center frequency tuning from 1.92 to 2.32 GHz under 3–30 V bias voltage, while maintaining 3-dB bandwidth at 440 ± 5 MHz, indicating exceptional bandwidth stability. Passband insertion loss remained low between 2.1 and 2.6 dB. The filter occupies 25.0 mm × 12.8 mm, validating the effectiveness of this area-saving topology, which demonstrates scalability for higher-order implementations.

We fabricated a sixth-order tunable bandpass filter using F4B-2 substrate (ε_r = 2.65, thickness 0.8 mm), implementing varactor diode (BB179) tuning through a DC bias path. To suppress RF leakage in the bias network, a DC feed circuit was constructed using two RF chip inductors (0603CS82N). Results in Fig. 3.25 show excellent agreement between measurements and simulations. The measured center frequency continuously tunes from 1.9 to 2.3 GHz while maintaining an absolute bandwidth of 580 ± 10 MHz, with insertion loss varying between 2.8 and 3.2 dB. Three distinct transmission zero pairs generated at band edges synchronously shift with tuning frequency, significantly enhancing out-of-band rejection. This design achieves constant absolute bandwidth in a sixth-order topology without auxiliary tuning components. The overall footprint measures 40 mm × 35 mm.

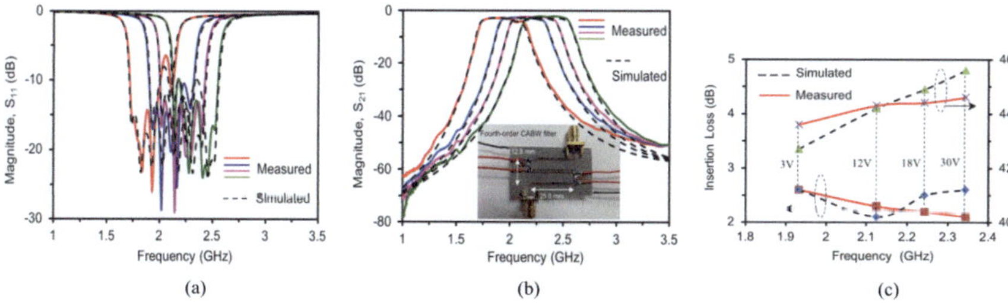

Fig. 3.24 Simulated and measured results. **a** S_{11}. **b** S_{21}. **c** Insertion loss and 3-dB bandwidth of fourth-order tunable filter for bias voltage of: 3, 12, 18, and 30 V (from left to right)

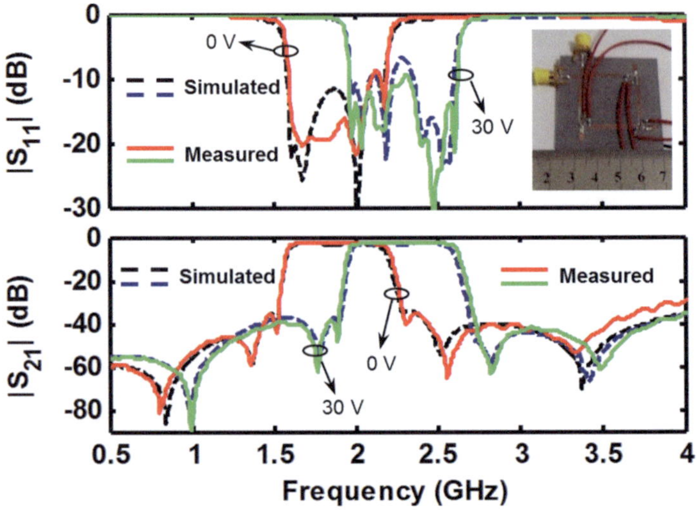

Fig. 3.25 Simulated and measured S-parameters of the proposed six-pole tunable filters for bias voltage of 0 and 30 V

3.4.2 Validation of Constant Fractional Bandwidth (CFBW) Tunable Filter

To validate the design theory for constant-fractional-bandwidth filters, we implemented a third-order filter based on cross-coupled cascaded tri-section (CT) topology as shown in Fig. 3.26a. Measured results in Fig. 3.26b demonstrate continuous center frequency tuning from 950 to 1460 MHz under varying bias voltages while maintaining constant fractional bandwidth (12.5% ± 2%). Insertion loss ranges from 1.8 to 3.6 dB with return loss exceeding 10 dB. By adjusting the voltage of cross-coupling capacitors, the transmission zero at the lower passband edge synchronously shifts with the tuning frequency, preserving steep roll-off characteristics. Electromagnetic simulations and measurements show high correlation in the lower passband region, with high-frequency discrepancies attributed to parasitic inductance and structural parasitic effects.

To further extend bandwidth adjustability, we developed a fourth-order filter using cascaded quad-section topology (Fig. 3.27a). Experimental results (Fig. 3.27b) show 50% fractional tuning range (860–1290 MHz center frequency) with maintained constant fractional bandwidth (15.9% ± 2.8%). Insertion loss varies between 2.6 and 3.2 dB while return loss remains better than 10 dB. Independent control of cross-coupling capacitor biases enables synchronous movement of the lower passband transmission zero with tuning frequency, consistently sustaining sharp roll-off characteristics. Strong agreement between simulated and measured passband performance validates the methodology accuracy.

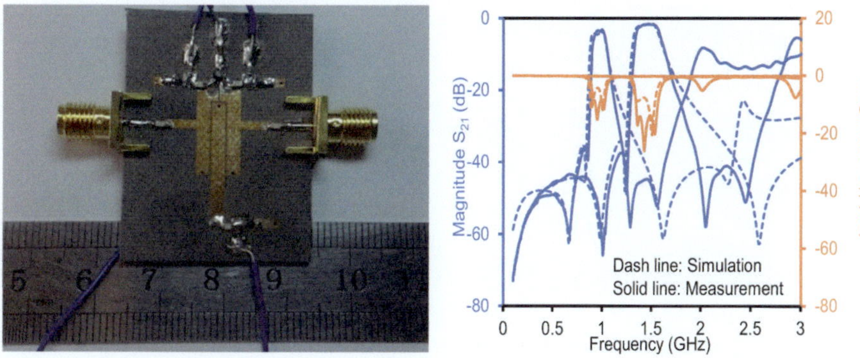

(a) (b)

Fig. 3.26 Results of the fabricated 3-pole tunable BPF based on CT topology. **a** Photograph. **b** Comparison between measurement and simulation

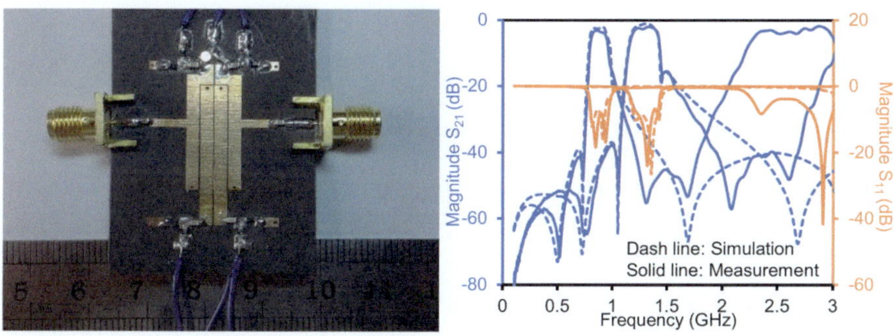

(a) (b)

Fig. 3.27 Results of the fabricated 4-pole tunable BPF based on CQ topology. **a** Photograph. **b** Comparison between measurement and simulation

3.4.3 Validation of Tunable Filter with Controllable/ Reconfigurable Bandwidth

To validate the proposed tunable-bandwidth filter design, we fabricated a microstrip prototype on F4B-2 substrate ($\varepsilon_r = 2.65$, thickness 0.5 mm) featuring a varactor-loaded triple-mode resonator structure, as depicted in Fig. 3.28a. Load capacitance C_L was implemented with chip capacitors (1.5 mm × 0.7 mm), utilizing abrupt-junction tuning varactors (SMV1405, SC-79 package from SKYWORKS) whose capacitance values measured 0.63 pF at 30 V bias and 2.67 pF at 0 V bias. DC biasing was achieved via a 100 kΩ resistor and ATC chip capacitor connected between the DC voltage source and the resonator's open end. By incorporating a series chip capacitor

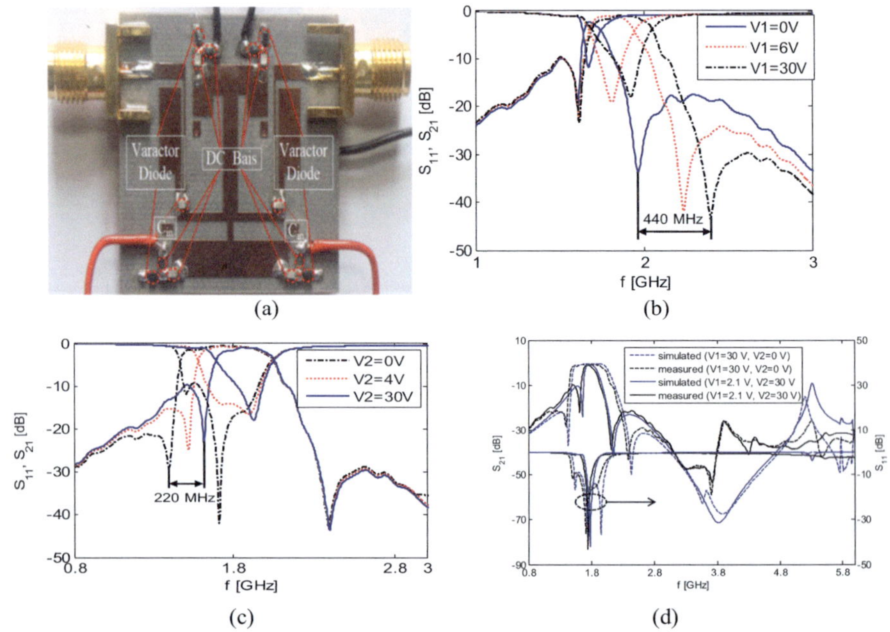

Fig. 3.28 Results of fabricated bandwidth tunable BPF with harmonic suppression. **a** Photograph. **b** Measured S-parameters for upper and lower bandage tuning: fixed $V_2 = 30$ V, varying V_1. **c** Measured S-parameters for upper and lower bandage tuning: fixed $V_1 = 30$ V, varying V_2. **d** Measured S-parameters for the bandwidth tunable BPF with harmonic suppression

$C_{dc} = 2.2$ pF with the varactor diodes, the net capacitance reduced to 0.49–1.20 pF. ATC 600S RF capacitors balanced tunability and loss.

Simulation and measured results in Fig. 3.28b–d show resonant frequencies shifting toward higher bands with increasing reverse bias voltage. As (b) demonstrates, with C_2 bias fixed at 30 V, TZ_1 tunes across 440 MHz (1.96–2.40 GHz); (c) indicates that with C_1 bias held at 30 V, TZ_2 tunes from 1.40 to 1.62 GHz; (d) exhibits the tunable-bandwidth filter performance: at a fixed 1.75 GHz center frequency, the − 3 dB bandwidth adjusts from 200 to 560 MHz (relative bandwidth 11.4–32.0%) with insertion loss below 1.1 dB, demonstrating exceptional simulation-measurement correlation.

Additionally, we designed and fabricated a tunable-bandwidth filter based on a quadruple-mode resonator, shown in Fig. 3.29a, manufactured on a 0.5-mm-thick F4B-2 substrate ($\varepsilon_r = 2.65$). Terminating capacitors C_1 and C_2 were implemented using integrated chip capacitors (1.8 mm × 1.8 mm) featuring SKYWORKS SMV-1405 abrupt-junction tuning varactor diodes in SC-70 packages. A common-cathode design enabled isolated dual DC biasing. Individual varactors exhibited capacitances of 0.63 pF at 30 V reverse bias and 2.67 pF at 0 V bias. Three 0603CS82N RF chip inductors effectively suppressed RF leakage in the bias network.

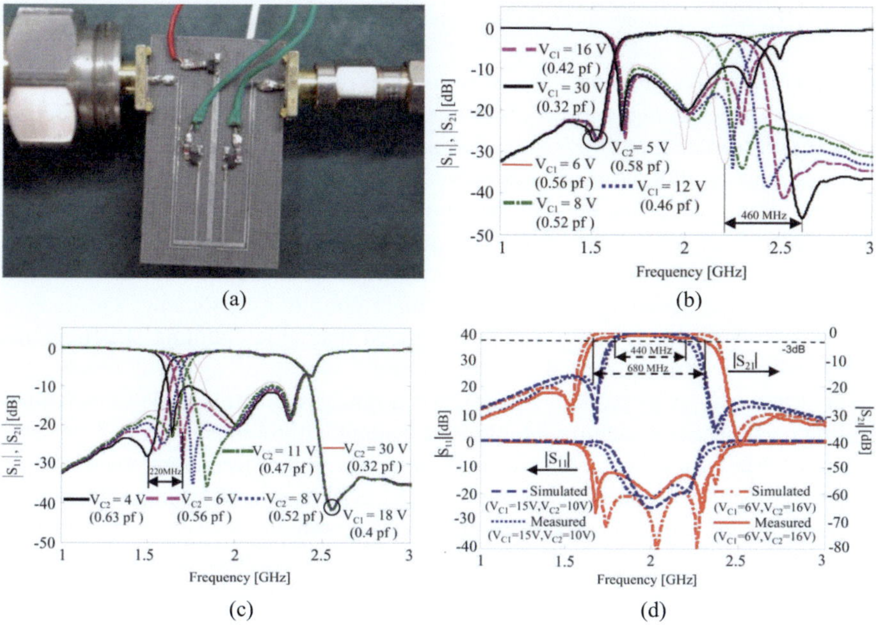

Fig. 3.29 Results of fabricated bandwidth tunable BPF with quadruple-mode resonator. **a** Photograph. **b** Measured S-parameters for two different cases: fixed $V_{C2} = 5$ V, varying V_{C1}. **c** Measured S-parameters for two different cases: fixed $V_{C1} = 18$ V, Varying V_{C2}. **d** Measured and simulated S-parameters of the filter with tunable bandwidth

Figure 3.29b–d demonstrate the tunability of resonant frequencies and passband characteristics. Increasing reverse voltage on varactors shifts corresponding resonant frequencies and bandwidths toward higher frequencies. When C_1 bias varies from 6 to 30 V with C_2 fixed at 5 V, TZ_1 tunes across 460 MHz (2.22–2.68 GHz). With C_2 voltage adjustment, TZ_2 tunes from 1.51 to 1.73 GHz. Final measurements demonstrate core bandwidth reconfiguration performance: At a fixed 2.0 GHz center frequency, the − 3 dB bandwidth continuously tunes from 440 to 680 MHz (fractional bandwidth: 22–34%), while return loss ($|S_{11}|$) remains better than − 12 dB across all tuning states. Measured results align closely with simulations, with minor discrepancies attributed to biasing circuit implementation and fabrication tolerances.

3.5 Conclusion

This chapter delves into the in-depth design and research of bandwidth-controllable bandpass filters. Addressing the bottleneck of traditional tunable filters where bandwidth and center frequency cannot be independently controlled, it proposes several innovative circuit topologies. The core solutions include: successfully realizing center-frequency tunable filters with constant absolute bandwidth characteristics by utilizing a dual open-loop resonator structure supporting hybrid electromagnetic coupling and implementing dynamic control of the external quality factor (Q_{ext}); further designing high-order constant absolute bandwidth filters from second-order to sixth-order, including even a compact square-layout sixth-order filter, based on topologies employing shunt-fed coupled lines (J-inverters) and alternating J/K inverters. Simultaneously, to meet the demands of modern communication systems for constant fractional bandwidth, techniques for achieving constant relative bandwidth are explored in cross-coupled cascaded triple (CT) and quadruplet microstrip filter topologies, ensuring stability in the main coupling coefficients and external quality factors during tuning through optimized magnetic coupling structures and port networks. Notably, based on decoupling analysis of resonant modes, two novel structures—a tri-mode resonator and a quad-mode stub-loaded resonator—are proposed. These structures enable independent, continuous tuning of the lower and upper band-edge frequencies while keeping the center frequency fixed, thereby dynamically adjusting the filter bandwidth and effectively controlling transmission zeros to enhance selectivity. Rigorous theoretical parameter analysis derives the key conditions and formulas essential for achieving constant bandwidth and bandwidth control. Finally, the feasibility and superiority of the proposed filter design schemes are comprehensively verified through simulation and measured data from multiple design examples ranging from second-order to sixth-order filters. The measured results demonstrate that these filters maintain constant absolute or fractional bandwidth while tuning the center frequency, or conversely, dynamically adjust the bandwidth while keeping the center frequency fixed, all while maintaining excellent insertion loss and selectivity. These bandwidth-controllable tunable filter design schemes provide an efficient and flexible core component solution for reconfigurable RF front-ends.

References

1. S. J. Park and G. M. Rebeiz, Low-loss two-pole tunable filters with three different predefined bandwidth characteristics, IEEE Trans Microwave Theory Tech 56 (2008), 1137–1148.
2. M. A. El-Tanani and G. M. Rebeiz, Corrugated microstrip coupled lines for constant absolute bandwidth tunable filters, IEEE Trans Microwave Theory Tech 58 (2010), 956–963.
3. M. A. El-Tanani and G. M. Rebeiz, High-performance 1.5–2.5-GHz RF-MEMS tunable filters for wireless applications, IEEE Trans Microwave Theory Tech 58 (2010), 1629–1637.
4. X. Y. Zhang and Q. Xue, High-selectivity tunable bandpass filters with harmonic suppression, IEEE Trans Microwave Theory Tech 58 (2010), 964–969.
5. X. Y. Zhang, Q. Xue, C. H. Chan and B. J. Hu, Low-loss frequency-agile bandpass filters with controllable bandwidth and suppressed second harmonic, IEEE Trans Microwave Theory Tech 58 (2010), 1557–1564.
6. Q. Y. Xiang, Q. Y. Feng, X. G. Huang and D. H. Jia, Electrical tunable microstrip LC bandpass filters with constant bandwidth, IEEE Trans Microwave Theory Tech 61 (2013), 1124–1130.
7. X. G. Wang, Y. H. Cho and S. W. Yun, A tunable combline bandpass filter loaded with series resonator, IEEE Trans Microwave Theory Tech 60 (2012), 1569–1576.

8. W. X. Tang and J. S. Hong, Varactor-tuned dual-mode bandpass filters, IEEE Trans Microwave Theory Tech 58 (2010), 2213–2219.
9. L. Athukorala and D. Budimir, Compact second-order highly linear varactor-tuned dual-mode filters with constant bandwidth, IEEE Trans Microwave Theory Tech 59 (2011), 2214–2220.
10. K. Chan Ho and C. Kai, Ring resonator bandpass filter with switchable bandwidth using stepped-impedance stubs, IEEE Trans Microwave Theory Tech, 58 (2010), 3936–3944.
11. X. Huang, Q. Y. Feng, Q. Y. Xiang, Bandpass filter with tunable bandwidth using quadruple-mode stub-loaded resonator, IEEE Microwave and Wireless Components Letters, 23(2012), 176–178.
12. W. Y. Park and S. Lim, Bandwidth tunable and compact band-pass filter (BPF) using complementary split ring resonators (CSRRS) on substrate integrated waveguide (SIW), Journal of Electromagnetic Waves and Applications, 24(2010), 2407–2417.
13. A. Lacorte Caniato Serrano, F. Salete Correra, T. P. Vuong and P. Ferrari, "Synthesis methodology applied to a tunable patch filter with independent frequency and bandwidth control," IEEE Transactions on Microwave Theory and Techniques, vol. 60, pp. 484–493, 2012.
14. S. J. Park, K. Y. Lee and G. M. Rebeiz, "Low-loss 5.15–5.70-GHz RF MEMS switchable filter for wireless LAN applications," IEEE Transactions on Microwave Theory and Techniques, vol. 54, pp. 3931–3939, 2006.
15. C. Yi-Chyun and G. M. Rebeiz, "A quasi elliptic function 1.75–2.25 GHz 3-pole bandpass filter with bandwidth control," IEEE Transactions on Microwave Theory and Techniques, vol. 60, pp. 244–249, 2012.
16. B. Liu, F. Wei, H. Zhang, X. Shi, "A tunable bandpass filter with switchable bandwidth," Journal of Electromagnetic Waves and Applications, vol. 25, pp. 223–232, 2011.
17. Chen. C.F., Huang T.Y. and Wu R.B., "Design of dual- and triple-passband filters using alternately cascaded multiband resonators," IEEE Transactions on Microwave Theory and Techniques, vol. 54, pp. 3550–3558, 2006.
18. Tang C.W and Wu P.H., "Design of a planar dual-band bandpass filter," IEEE Microwave and Wireless Components Letters, vol. 21, pp. 362–364, 2011.
19. Zhou X.J., Zhao Y.J., Fu Y. and Liu Y.Y., "Compact dual-mode tri-band microstrip BPF with three sets of resonators," Progress In Electromagnetics Research Letters, vol. 33, pp. 47–54, 2012.
20. Y. C. Chiou and G. M. Rebeiz, A tunable three-pole 1.5–2.2-GHz bandpass filter with bandwidth and transmission zero control, IEEE Trans Microwave Theory Tech 59 (2011), 2872–2878.
21. J. R. Mao, W. W. Choi, K. W. Tam, W. Q. Che and Q. Xue, Tunable bandpass filter design based on external quality factor tuning and multiple mode resonators for wideband applications, IEEE Trans Microwave Theory Tech 61 (2013), 2574–2584.
22. Y. C. Chiou and G. M. Rebeiz, Tunable 1.55–2.1 GHz 4-pole elliptic bandpass filter with bandwidth control and > 50 dB rejection for wireless systems, IEEE Trans Microwave Theory Tech 61 (2013), 117–124.
23. Huang, X.-G., Feng, Q.-Y., Xiang, Q.-Y. and Jia, D.-H. (2014), Constant absolute bandwidth tunable filter using varactor-loaded open-loop resonators. Microw. Opt. Technol. Lett., 56: 1178–1181. https://doi.org/10.1002/mop.28285.
24. Huang, X., Feng, Q., Zhu, L., Xiang, Q. and Jia, D. (2014), Synthesis and design of tunable bandpass filters with constant absolute bandwidth using varactor-loaded microstrip resonator. Int J RF and Microwave Comp Aid Eng, 24: 681–689. https://doi.org/10.1002/mmce.20812.
25. X.-G. Huang, Jin-Qi Zhang, Yi-Qun Lin and Q.-Y. Xiang, "Design of a six-pole tunable bandpass filter with constant absolute bandwidth," 2016 Progress in Electromagnetic Research Symposium (PIERS), Shanghai, China, 2016, pp. 3507–3510, https://doi.org/10.1109/PIERS.2016.7735357.
26. Jia, D., Feng, Q., Huang, X., & Xiang, Q. (2017). Tunable microstrip bandpass filter with constant fractional band-width based on cascade triplet topology. International Journal of Electronics, 104(10), 1646–1657. https://doi.org/10.1080/00207217.2017.1321142.
27. D.-H. Jia and J. Deng, "Tunable microstrip bandpass filter based on cascade quadruplet topology," 2017 Progress in Electromagnetics Research Symposium - Fall (PIERS - FALL), Singapore, 2017, pp. 777–779, https://doi.org/10.1109/PIERS-FALL.2017.8293240.

28. Ding-Hong Jia, Quanyuan Feng, Xiao-Guo Huang, and Qian-Yin Xiang, "A Novel Tri-Mode Bandwidth Tunable Filter with Harmonic Suppression," Progress In Electromagnetics Research C, Vol. 34, 183–194, 2013. https://doi.org/10.2528/PIERC12092503.
29. X. Huang, Q. Feng and Q. Xiang, "Bandpass Filter With Tunable Bandwidth Using Quadruple-Mode Stub-Loaded Resonator," in IEEE Microwave and Wireless Components Letters, vol. 22, no. 4, pp. 176–178, April 2012, https://doi.org/10.1109/LMWC.2012.2188280.
30. L. Zhu and K. Wu, Accurate circuit model of interdigital capacitor and its application to design of new quasi-lumped miniaturized filters with suppression of harmonic resonance, IEEE Trans Microwave Theory Tech 48 (2000), 347–356.
31. S. B. Zhang and L. Zhu, Synthesis method for even-order symmetrical Chebyshev bandpass filters with alternative J/K inverters and λ/4 resonators, IEEE Trans Microwave Theory Tech 61 (2013), 808–816.
32. J. S. Hong and M. Lancaster, Microstrip Filters for RF/Microwave Applications. New York: Wiley, 2001.

Chapter 4
Tunable Filter with Controllable Multi-bands

4.1 Introduction

In previous chapters, we analyzed the design theory of highly selective bandpass filters based on multimode resonators and constant-bandwidth tunable bandpass filters. With the rapid development of modern communication systems, radar, and electronic warfare systems, tunable multi-band bandpass filters [1–4], driven by the demand for multi-band operation, have become an active research area.

Existing studies, such as [5–7], have primarily focused on designs where the first passband is fixed and the second passband is tunable. Less attention has been paid to independent dual-band tuning. Reference [8] demonstrated a varactor-diode-tuned dual-band tunable bandpass filter with independently tunable passbands. Yet, this filter occupies a substantial circuit size, requires a large number of varactor diodes, and makes the bias circuit highly complex. In [9], the authors proposed a tunable dual-band bandpass filter based on two varactor-diode-tuned resonators. This structure offers the possibility of achieving two tunable passbands: either with a fixed first passband and a controllable second passband, or with both passbands tuned concurrently. Nevertheless, the two passbands in this design cannot achieve independent tuning. Moreover, due to the coupled excitation of the odd- and even-mode resonances of the employed dual-mode resonator, the two passbands significantly influence each other. Subsequently, [10] implemented a dual-band filter with independently tunable dual passbands using two dual-mode single-band filters sharing input/output lines. However, this filter requires the application of four different bias voltages to control the center frequency and bandwidth, introducing control complexity.

In summary, these reported tunable dual-band filters still face numerous unresolved challenges. The first challenge is the lack of true independence: some works can only achieve tunability for one passband while keeping the other fixed, or exhibit strong coupling between the two modes/bands. This coupling inevitably causes tuning one passband's characteristics to affect the other, preventing genuine independent tuning. The second challenge is structural complexity. Conventional tunable

F. Nian et al., *RF & Microwave Multi-mode and Tunabe Filters*,
https://doi.org/10.1007/978-981-92-0577-6_4

dual-band bandpass filter structures aiming for independent tuning of each passband's center frequency and bandwidth often necessitate introducing separate resonant structures, tuning elements, coupling paths, and bias control networks for each passband. Consequently, this leads to a significant increase in the physical structure's complexity and size, tunable circuitry complexity, insertion loss, and other performance compromises.

This chapter will explore two core technical approaches for implementing bandpass filters featuring independently tunable dual passbands. The primary objective is to dynamically generate two passbands simultaneously within a single filter structure, enabling independent control over both the center frequency and bandwidth of each passband. The approach based on generating independent modes for dual-band adjustment centers on controlling multiple internal modes via resonators to form tunable passbands. Through specific structural designs, such as loading tuning elements (e.g., varactor diodes) at strategic positions, independent regulation of the resonant frequency for different modes is achieved. The second approach achieves multi-band adjustment by constructing distinct signal transmission paths. Its core lies in utilizing multiple resonators to form mutually isolated signal transmission paths. Each path is individually controlled via its own tuning elements, and optimized design minimizes crosstalk between paths, enabling the creation of multiple separately tunable passbands. Based on this, the subsequent sections of this chapter will proceed as follows: Sect. 4.2 will provide an in-depth analysis of the operating principles of the aforementioned two implementation paths. Section 4.3 will introduce the design methodologies for specific filter structures based on these paths, present their simulated and measured results, evaluate performance metrics, and provide systematic theoretical guidance and engineering practice references for designing high-performance, independently controllable multi-band tunable filters. Finally, Sect. 4.4 will present a comprehensive summary of the chapter's content.

4.2 Theoretical Analysis

4.2.1 Based on Creating Independent Modes for Dual-Band Tuning

This method leverages the characteristic of dual-mode or multimode resonators to simultaneously support multiple distinct resonant modes (e.g., even and odd modes). The theoretical foundation is analyzed below using a dual-mode resonator as an example.

Odd- and Even-Mode Excitation Analysis

The configuration of the filter structure is based on two coupled center-loaded split-ring tunable resonators. The filter uses a modified dual-mode resonator with the loading varactor at the center of a transmission line and the input feed points are located symmetrically at the center so as to create a desired transmission zero between the two passbands [11].

Figure 4.1a shows the basic schematic of this dual-mode resonator, where four loading capacitances are attached at the resonator. C_{odd} is placed between the split of the loop, two identical capacitances C_1 are loaded at the ends of a transmission line with characteristic admittance Y_1 and physical length L_1, and C_{even} is added at the center of a transmission line, which is located at the plane of symmetry. For simplicity in analysis, all the transmission lines are assumed to be lossless and parasitic effects of the used varactor diodes are neglected. As shown in Fig. 4.1a, the physical geometry of the resonator is symmetrical, so the well-known odd- and even-mode method can be adopted to analyze this structure. Under resonance, this resonator should achieve zero input admittance for both the odd and even modes, i.e.,

$$\mathrm{Im}[Y_{ino}] = \mathrm{Im}[Y_{ine}] = 0 \tag{4.1}$$

For the odd-mode case, the plane of symmetry is the perfect electric wall or ideal short-circuit terminal in equivalence. As such, the whole resonator can be represented by its half a bisection circuit, as given in Fig. 4.1b. The odd-mode input admittance is derived as

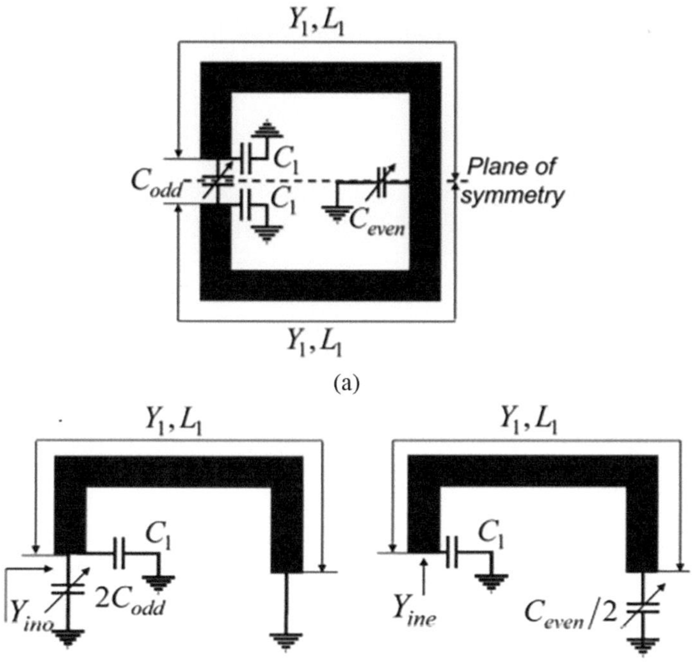

Fig. 4.1 Schematic of dual-mode resonator. **a** Whole schematic. **b** Equivalent odd-mode circuit. **c** Equivalent even-mode circuit

$$Y_{ino} = j\left(\omega_o C_1 + 2\omega_o C_{odd} - \frac{Y_1}{\tan(\beta_o L_1)}\right) \tag{4.2}$$

where β_o represents the phase constant at the odd-mode resonant angular frequency ω_o. Using (4.1) and (4.2), the odd-mode resonant frequency can be determined as

$$f_{odd} \times \tan\left(\frac{2\pi f_{odd} L_1 \sqrt{\varepsilon_r}}{v_0}\right) = \frac{Y_1}{2\pi C_1 + 4\pi C_{odd}} \tag{4.3}$$

where v_0 and ε_r are the speed of the light in free space and permittivity of a dielectric substrate.

The fundamental odd-mode resonant mode is used for forming the first passband. From (4.3), we can well understand that the odd-mode resonant frequency primarily depends on the loaded capacitances C_{odd} and C_1, whereas it is almost not affected by the loaded capacitance C_{even}. Therefore, the variation in C_{odd} or C_1 results in tuning this odd-mode resonant frequency of the proposed resonator in Fig. 4.1, enabling the first passband to be tunable in its central frequency. Moreover, when C_1 is fixed, the odd-mode resonant frequency fully depends on the loaded capacitance C_{odd} at the split of the ring.

For the even-mode case, the plane of symmetry becomes the perfect magnetic wall or ideal open-circuited terminal in equivalence. Thus, the even-mode input admittance, in its bisection circuit shown in Fig. 4.1c, is given

$$Y_{ine} = j\left(\omega C_1 + Y_1 \frac{\omega_e C_{even}/2 + Y_1 \tan(\beta_e L_1)}{Y_1 - \omega_e C_{even}/2 \tan(\beta_e L_1)}\right) \tag{4.4}$$

where β_e is the phase constant at the even-mode resonant angular frequency. Using (4.1) and (4.4), the even-mode resonant frequency can be found

$$\left(f_{even} - \frac{Y_1^2}{2\pi^2 f_{even} C_1 C_{even}}\right) \tan\left(\frac{2\pi f_{even} L_1 \sqrt{\varepsilon_r}}{v_0}\right) = \frac{Y_1\left(C_1 + \frac{C_{even}}{2}\right)}{\pi C_1 C_{even}} \tag{4.5}$$

In this aspect, the fundamental even-mode resonant mode will be used to construct the second passband. From (4.5), it is observed that the even-mode resonant frequency mainly depends on C_{even} and C_1, whereas hardly affected by C_{odd}. Thus, the variation in C_{even} or C_1 results to change the respective resonant frequency, enabling the tunability of the second passband frequency. When C_1 is fixed, the even-mode resonant frequency fully depends on C_{even}, which is located at the center of a transmission line resonator. Based on the above discussion, the two tunable dual resonant modes coexisting in the proposed resonator have an attractive capability in design of a novel dual-band BPF with independently controllable dual passbands using these two resonant modes. Moreover, the tuning range in dual passbands can be extended by loading a capacitance C_1.

To verify the above analysis, full-wave simulation is conducted to derive the frequency response of the proposed filter structure. To excite the resonator, two

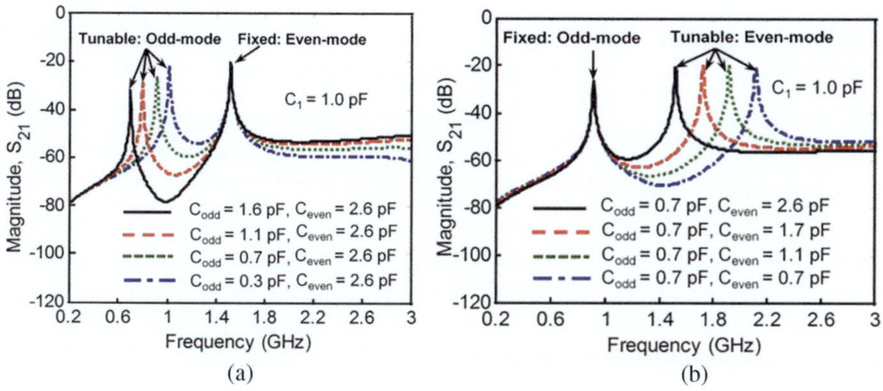

Fig. 4.2 Graphical demonstration of even- or odd-mode resonant frequencies versus two distinct capacitances. **a** Varied odd-mode capacitance (C_{odd}). **b** Varied even-mode capacitance (C_{even})

microstrip lines with 50 Ω are weakly coupled to the resonator. The simulated S_{21}-magnitude results are plotted in Fig. 4.2a, b for the different values of C_{odd} and C_{even}. The capacitor C_1 and length L_1 of the resonator are fixed at 1.0 pF and 30.0 mm, respectively.

Figure 4.2a shows that under the fixed even-mode resonant frequency, the odd-mode resonant frequency can be tuned by varying the capacitance of the varactor diode which connected two ends of the line. By varying the value of C_{odd} from 1.6 to 0.3 pF, the odd-mode resonant frequency increases from 0.67 to 1.0 GHz. This property can be used to design a dual-band BPF with the fixed second passband and controllable first passband.

Figure 4.2b shows that under the fixed odd-mode resonant frequency, the even-mode resonant frequency can be tuned by varying the capacitance of the varactor diode. By varying C_{even} from 2.6 to 0.7 pF, the odd-mode resonant frequency can be enlarged from 1.5 to 2.1 GHz. This characteristic can be utilized to design a dual-band BPF with the fixed first passband and controllable second passband.

Coupling Coefficient (K_i)

To determine the coupling coefficients as inquired in design of a filter with specified bandwidth, the input admittances need to be deduced. For the balanced-type second-order dual-band tunable filter in Fig. 4.3, its respective coupling coefficient can be calculated by the odd- and even-mode equivalent circuits [12] as shown in Fig. 4.4a, b, respectively. In this way, the input admittances Y_{in_o1} and Y_{in_o2} are

$$Y_{in\_o1} = Y_1 \frac{\left(Y_S + Y_{Load\_o1}\right) + jY_1 \tan\theta_1}{Y_1 + \left(Y_S + Y_{Load\_o1}\right)\cot\theta_2 \tan\theta_1} + j\omega(2C_{odd} + C_1) \tag{4.6}$$

$$Y_{in\_o2} = Y_1 \frac{\left(Y_S + Y_{Load\_e1}\right) + jY_1 \tan\theta_1}{Y_1 + \left(Y_S + Y_{Load\_e1}\right)\cot\theta_2 \tan\theta_1} + j\omega(2C_{odd} + C_1) \tag{4.7}$$

where

$$Y_S = \frac{j\omega C_{in}(Y_0 + j\omega C_M)}{j\omega C_{in} + Y_0 + j\omega C_M} \tag{4.8}$$

$$Y_{load\_o1,e1} = Y_2 \frac{-jY_{3o,3e}\cot\theta_{3o,3e} + jY_2\tan\theta_2}{Y_2 + Y_{3o,3e}\cot\theta_{3o,3e}\tan\theta_2} \tag{4.9}$$

where Y_0 is the characteristic impedance of input/out feed line. Y_{3o} and Y_{3e} and θ_{3o} and θ_{3e} are the impedance and electric length of the coupling-line, respectively. Y_1 and Y_2 and θ_1 and θ_2 are the impedance and electric length of the transmission lines divided by the input/output tap point.

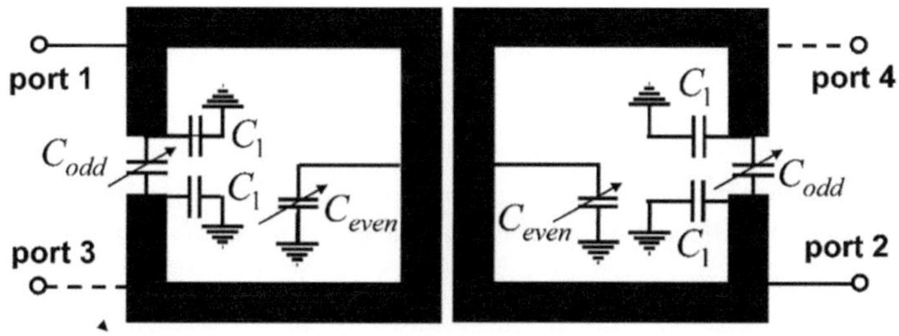

Fig. 4.3 Schematic of the two coupled resonators with a pair of balanced ports

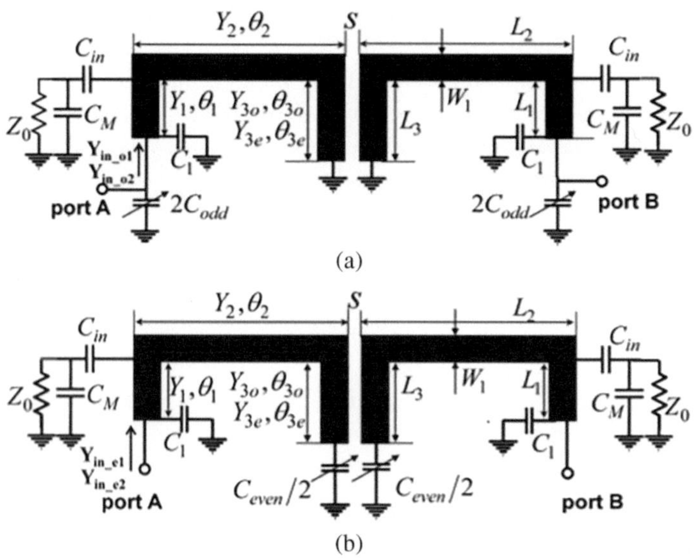

Fig. 4.4 Equivalent circuits for determinant of coupling coefficients under odd- and even-mode excitation. **a** Odd-mode bisection. **b** Even-mode bisection. $L_1 = 9.0$ mm, $L_2 = 10.0$ mm, $W_1 = 1.0$ mm, $C_1 = 0.6$ pF, $C_{in} = 15.0$ pF, $C_M = 1.5$ pF and $Z_0 = 50\ \Omega$

As such, the resonant frequencies can be calculated by

$$\mathrm{Im}\left[Y_{in\_o1}(\omega_{o1})\right] = \mathrm{Im}\left[Y_{in\_o2}(\omega_{02})\right] = 0 \tag{4.10}$$

Meanwhile, the coupling coefficient K_1 of the first passband can be calculated by the following formulas as well studied [13]:

$$K_1 \approx \frac{\omega_{o2}^2 - \omega_{o1}^2}{\omega_{o2}^2 + \omega_{o1}^2} \tag{4.11}$$

For the even-mode equivalent circuit in Fig. 4.4b, the input admittances Y_{in_e1} and Y_{in_e2} are

$$Y_{in\_e1} = Y_1 \frac{\left(Y_S + Y_{Load\_o2}\right) + jY_1 \tan\theta_1}{Y_1 + \left(Y_S + Y_{Load\_o2}\right)\cot\theta_2 \tan\theta_1} + j\omega C_1 \tag{4.12}$$

$$Y_{in\_e2} = Y_1 \frac{\left(Y_S + Y_{Load\_e2}\right) + jY_1 \tan\theta_1}{Y_1 + \left(Y_S + Y_{Load\_e2}\right)\cot\theta_2 \tan\theta_1} + j\omega C_1 \tag{4.13}$$

where

$$Y_{load\_o2,e2} = Y_2 \frac{jY_{3o,3e}\frac{\omega C_{even} + Y_{3o,3e}\tan\theta_{3o,3e}}{Y_{3o,3e} - \omega C_{even}\tan\theta_{3o,3e}} + jY_2 \tan\theta_2}{Y_2 - jY_{3o,3e}\frac{\omega C_{even} + Y_{3o,3e}\tan\theta_{3o,3e}}{Y_{3o,3e} - \omega C_{even}\tan\theta_{3o,3e}}\tan\theta_2} \tag{4.14}$$

Similarly, the coupling coefficient in Fig. 4.4b can be solved by

$$\mathrm{Im}\left[Y_{in\_e1}(\omega_{e1})\right] = \mathrm{Im}\left[Y_{in\_e2}(\omega_{e2})\right] = 0 \tag{4.15}$$

and the coupling coefficient K_2 of the second passband is

$$K_2 \approx \frac{\omega_{e1}^2 - \omega_{e2}^2}{\omega_{e1}^2 + \omega_{e2}^2} \tag{4.16}$$

Figure 4.5a, b plot the calculated coupling coefficients using (4.11) and (4.16) for the first and second passbands under different L_3 and S which are length and gap width of the coupling line, respectively. As L_3 increases, the coupling coefficient K_1 increases, whereas K_2 decreases. On the other hand, the slope of all the coupling coefficients is almost unchanged regardless of how L_3 varies. It also can be seen that the slope of the coupling coefficients seems to be changed with respect to S. As a result, L_3 and S can be determined using these design graphs to meet the requirement in the desired coupling coefficients in a designed filter with increasing fractional bandwidth at first passband and constant absolute bandwidth at second passband.

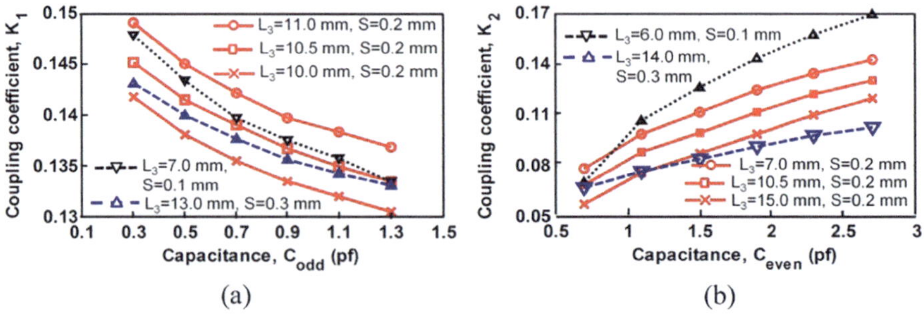

Fig. 4.5 Calculated coupling coefficient versus odd- and even-mode capacitances. **a** For the first passband. **b** For the second passband

External Quality Factor

Figure 4.6a, b depict the equivalent one-port odd- and even-mode circuit model with the I/O coupling to the resonator by capacitance C_{in} and C_M, respectively. The external quality factors, Q_{exo} and Q_{exe}, for the odd- and even-mode circuits can be derived from the input reflection coefficients as done in [14].

$$Q_{exo} = \pi f_{odd} \tau_{S11\_odd} \tag{4.17}$$

$$Q_{exe} = \pi f_{even} \tau_{S11\_even} \tag{4.18}$$

where τ_{S11_odd} and τ_{S11_even} are the group delays of S_{11} at the odd- and even-modes resonant frequencies, respectively.

To maintain the unchanged shape and specified bandwidth of the odd- and even-mode frequency responses over the desired tuning range, Q_{exo} and Q_{exe} of the proposed tunable filter must be satisfied with

$$Q_{exo} = \frac{g_0 g_1}{FBW_1} \tag{4.19}$$

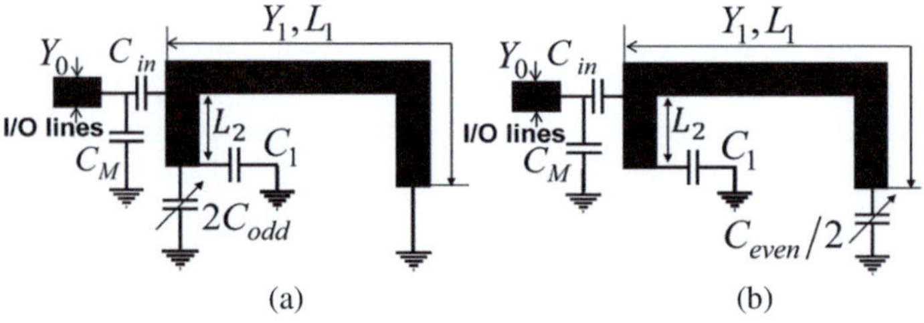

Fig. 4.6 Equivalent circuits with distinct I/O couplings. **a** Odd-mode case. **b** Even-mode case

$$Q_{exe} = \frac{g_0 g_1}{FBW_2} \tag{4.20}$$

where FBW_1 and FBW_2 represent the fractional bandwidth of the first and second passbands, respectively. g_0 and g_1 are the element values of the prototype lowpass filter. Equations (4.19) and (4.20) can be simultaneously achieved by properly choosing the input matching capacitance parameters C_{in}, C_M and tap point as shown in Fig. 4.6.

In summary, this approach primarily achieves dual-band functionality by strategically positioning tuning elements at distinct locations, thereby enabling each resonant mode to be independently governed by specific tuning components and thus forming separate passbands.

4.2.2 Based on Independent Signal Paths for Dual-Band Tuning

This approach channels the input signal into distinct resonant paths. Each path contains its own resonant unit and tuning elements. Through meticulous structural design, signals propagating along the two paths exhibit negligible mutual interference during tuning and operation, thereby enabling independent control of the two passbands. When analyzing such dual-band bandpass filter structures, the system can be modeled as two independent filters to examine the generation mechanism of each passband and transmission zero. The theoretical foundation is analyzed using the following two filter structures as examples:

Filter I

The dual-band characteristics is realized by two independent bandpass filters which consist of Filter I-a and Filter I-b. Filter I-a Consists of an input/output coupling structure and two quarter-wavelength varactor-loaded open-loop resonators [15]. For the two order BPF as shown in Fig. 4.7, the input admittance seen from port A and port B can be defined as

$$Y_{in\_o,e} = \Delta Y_{o,e} / Y22_{o,e} \tag{4.21}$$

$$Y_{o,e} = \begin{bmatrix} D_{o,e}/B_{o,e} + j\omega C_L & -\Delta A/B_{o,e} - j\omega C_L \\ -1/B_{o,e} - j\omega C_L & -A_{o,e}/B_{o,e} + j\omega C_L \end{bmatrix} \tag{4.22}$$

$$\Delta A = A_{o,e} D_{o,e} - B_{o,e} C_{o,e} \tag{4.23}$$

$$A = A_1 \cdot A_{2o,2e} \cdot A_{3o,3e} \cdot A_4 \tag{4.24}$$

$$A_{1,4} = \begin{bmatrix} \cos\phi_{1,4} & j\sin\phi_{1,4}/Y_{1,4} \\ j\sin\phi_{1,4}Y_{1,4} & \cos\phi_{1,4} \end{bmatrix} \tag{4.25}$$

$$A_{2o,e} = \begin{bmatrix} \cos\phi_{2o,e} & j\sin\phi_{2o,e}/Y_{2o,e} \\ j\sin\phi_{2o,e}Y_{2o,e} & \cos\phi_{2o,e} \end{bmatrix} \tag{4.26}$$

$$A_{3o,e} = \begin{bmatrix} X_{11} & X_{12} \\ X_{21} & X_{22} \end{bmatrix} \tag{4.27}$$

where

$$X_{11} = \frac{2(\cos\phi_{3o}\cos\phi_{3e} + 1) - \sin\phi_{3o}\sin\phi_e(Y_{3o}/Y_{3e} + Y_{3e}/Y_{3o})}{2(\cos\phi_{3o} + \cos\phi_{3e})} \tag{4.28}$$

$$X_{12} = j\frac{\sin\phi_{3o}\cos\phi_{3e}/Y_{3o} + \cos\phi_{3o}\sin\phi_{3e}/Y_{3e}}{\cos\phi_{3o} + \cos\phi_{3e}} \tag{4.29}$$

$$X_{21} = j\frac{\sin\phi_{3o}\cos\phi_{3e}Y_{3o} + \cos\phi_{3o}\sin\phi_{3e}Y_{3e}}{\cos\phi_{3o} + \cos\phi_{3e}} \tag{4.30}$$

$$X_{22} = \frac{2\cos\phi_{3o}\cos\phi_{3e}}{\cos\phi_{3o} + \cos\phi_{3e}} \tag{4.31}$$

The overall admittance matrix of capacitively loaded coupled resonator is

$$[Y] = \begin{bmatrix} \frac{Y_{in\_e}+Y_{in\_o}}{2} & \frac{Y_{in\_e}-Y_{in\_o}}{2} \\ \frac{Y_{in\_e}-Y_{in\_o}}{2} & \frac{Y_{in\_e}+Y_{in\_o}}{2} \end{bmatrix} \tag{4.32}$$

For the above overall admittance matrix, three conditions must be satisfied. One is the resonance condition and the others are the coupling condition, Q_{ext}. The conditions are

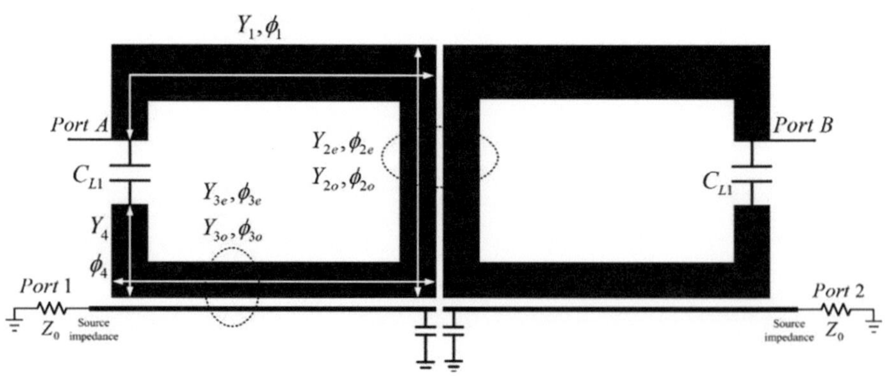

Fig. 4.7 Layout of Filter I-a

$$\mathrm{Im}(Y_{11}(\omega_0)) = 0 \quad \frac{\mathrm{Im}(Y_{12}(\omega_0))}{b} = K_{12} \quad Q_{ext} = \frac{b}{\mathrm{Re}[Y_{11}(\omega_0)]} = \frac{g_0 g_1}{\Delta} \tag{4.33}$$

where

$$b = \frac{\omega_0}{2}\frac{\partial \mathrm{Im}(Y_{11}(\omega_0))}{\partial \omega} \quad K_{12} = \frac{\Delta}{\sqrt{g_1 g_2}} \tag{4.34}$$

For simplicity and convenience to confirm the design parameters Y_1, $Y_{2e,o}$, $Y_{3e,o}$, $Y_4, w_1, w_2, w_3, l_1, l_2, l_3$, the loading capacitor C_{L1} can be decoupled from $\mathrm{Im}(Y_{11}(\omega_0)) = 0$, then the C_{L1} can be rewritten as

$$C_{L1} = \frac{1}{j\omega}\mathrm{Im}\left[\frac{C_o + C_e}{A_o + A_e + B_o D_o + B_e D_e - \Delta A_o - \Delta A_e - 2}\right]\bigg|_{\omega=\omega_0} \tag{4.35}$$

With the C_{L1} and the above equations, the other design parameters can be determined. Using the aforementioned method, the tunability of Filter I-a are shown in Fig. 4.8. The load capacitors C_{L1} are implemented with lump chip capacitors (1.5 × 0.7 mm^2), where SMV 1405 abrupt junction tuning varactors of SKYWORKS in SC-79 package have been used as the tuning elements. The single capacitance is 0.63 pF and 2.67 pF at 30 V and 0 V bias, respectively. The SPICE Model of the SMV1405 is adopted in the simulation.

Filter I-b consists of an input/output coupling structure and a half-wavelength resonator, as shown in Fig. 4.9, is mainly controlled by the length ($2l_4 + l_5$) of the half-wavelength resonator, as the formula (4.36) shown. Q_{ext} determines the bandwidth of the passband. The method to extract the Q_{ext} is shown in [15].

$$f_c = \frac{c}{2(2l_4 + l_5)\sqrt{\varepsilon_e}} \tag{4.36}$$

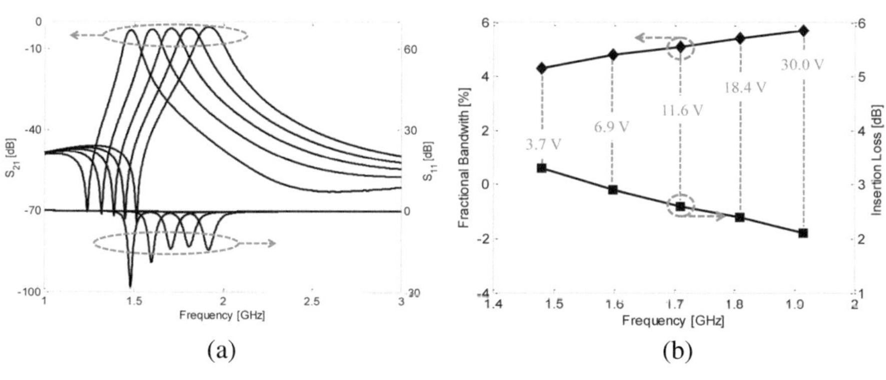

Fig. 4.8 Simulated results of Filter I-a. **a** Passband tunability. **b** Simulated insertion loss and 3-dB fractional bandwidth under the bias voltage $v1$ ($s1 = 0.1$ mm)

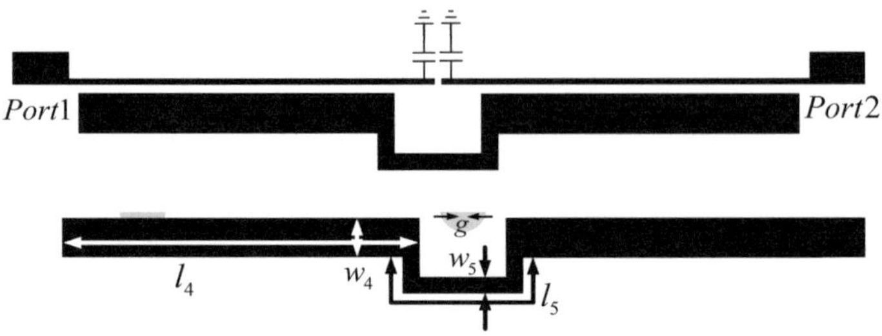

Fig. 4.9 Layout of Filter I-b

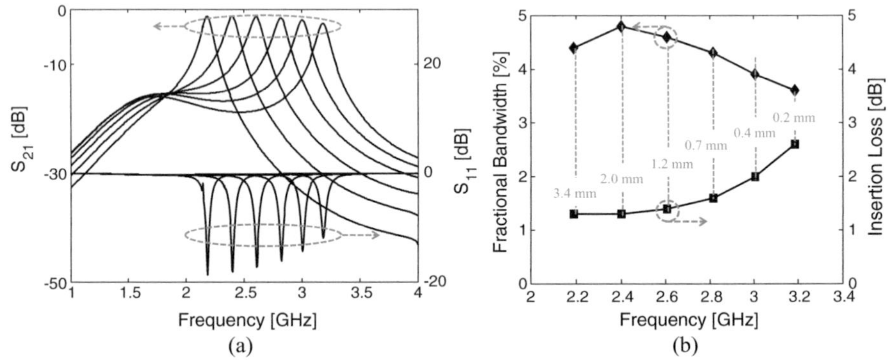

Fig. 4.10 Simulated results of Filter I-b. **a** Passband tunability. **d** Simulated insertion loss and 3-dB absolute bandwidth under the width w_4 ($s2 = 0.1$ mm)

Using the aforementioned method, the tunability of Filter I-b are shown in Fig. 4.10. The passband can be tuned by the length of the resonator. The longer the length is, the lower the frequency becomes, which is consistent with Eq. (4.36).

Filter II

Filter II consists of an input/output coupling structure and two dual-mode resonators [16]. Both dual-mode resonators (Filter II-a and Filter II-b) can be independently coupled to the input and output, resulting in two independently controllable passbands. The frequency resonant frequencies and Q_{ext} of the two dual-mode resonators will be derived in the following part. For the dual-mode resonator Filter II-a in Fig. 4.11, C_{load1} is supposed to be loaded varactors. All the three C_{load1} are to be used with a single bias voltage V_1. As the structure of the resonator is symmetrical, the even- and odd-mode analysis method is applicable to obtain the resonant frequencies. The resonant frequencies of the odd and even mode in Filter II-a can be found from $\mathrm{Im}[Y_{ino}] = 0$, $\mathrm{Im}[Y_{ine}] = 0$. The odd- and even-modes can be expressed

$$Y_{ino\_a} = j\left(\omega C_{load1} - Y_{1\_a} \cot \theta_{1\_a}\right) \tag{4.37}$$

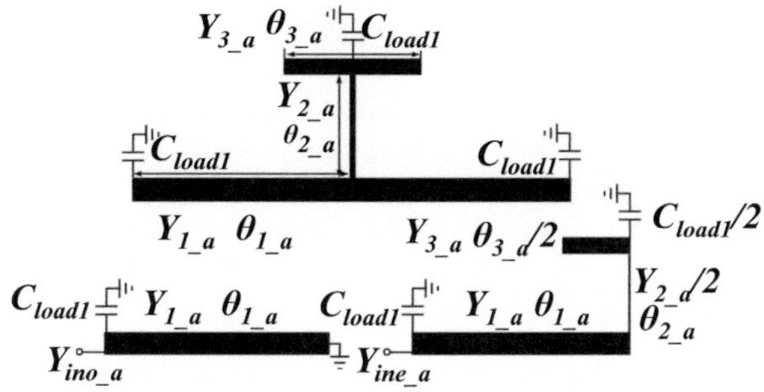

Fig. 4.11 Layout of Filter II-a and its even–odd-mode excitation equivalent circuit

$$Y_{ine\_a} = Y_{1\_a} \cdot \frac{Y_{Load1} + jY_{1\_a}\tan\theta_{1\_a}}{Y_{1\_a} + jY_{Load1}\tan\theta_{1\_a}} + j\omega C_{load1} \tag{4.38}$$

$$Y_{Load1} = \frac{Y_{2\_a}}{2} \cdot \frac{j\left(\omega C_{load1} + Y_{2\_a}\tan\theta_{2\_a} + 2Y_{3\_a}\tan(\theta_{3\_a}/2)\right)}{Y_{2\_a} - \left(\omega C_{load1} + 2Y_{3\_a}\tan(\theta_{3\_a}/2)\right)\tan\theta_{2\_a}} \tag{4.39}$$

For the dual-mode resonator Filter II-b in Fig. 4.12, C_{load2} is supposed to be loaded varactors. All the three C_{load2} are to be used with a single bias voltage V_2. Similar to Filter II-b, the odd- and even-modes of Filter II-b can be expressed

$$Y_{ino\_b} = jY_{1\_b} \cdot \frac{Y_{1\_b}\tan\theta_{1\_b} - Y_{2\_b}\cot\theta_{2\_b}}{Y_{1\_b} + Y_{2\_b}\cot\theta_{2\_b}\tan\theta_{1\_b}} + j\omega C_{load2} \tag{4.40}$$

$$Y_{ine\_b} = Y_{1\_b} \cdot \frac{Y_{2\_b} \cdot \frac{Y_{load2} + jY_{2\_b}\tan\theta_{2\_b}}{Y_{2\_b} + jY_{load2}\tan\theta_{2\_b}} + jY_{1\_b}\tan\theta_{1\_b}}{Y_{1\_b} + jY_{2\_b} \cdot \frac{Y_{load2} + jY_{2\_b}\tan\theta_{2\_b}}{Y_{2\_b} + jY_{load2}\tan\theta_{2\_b}}\tan\theta_{1\_b}} + j\omega C_{load2} \tag{4.41}$$

$$Y_{load2} = \frac{Y_{3\_b}}{2} \cdot \frac{\frac{jY_{4\_b}}{2} \cdot \frac{\omega C_{load2} + Y_{4\_b}\tan\theta_{4\_b}}{Y_{4\_b} - \omega C_{load2}\tan\theta_{4\_b}} + j\frac{Y_{3\_b}}{2}\tan\theta_{3\_b}}{\frac{Y_{3\_b}}{2} - \frac{Y_{4\_b}}{2} \cdot \frac{\omega C_{load2} + Y_{4\_b}\tan\theta_{4\_b}}{Y_{4\_b} - \omega C_{load2}\tan\theta_{4\_b}}\tan\theta_{3\_b}} \tag{4.42}$$

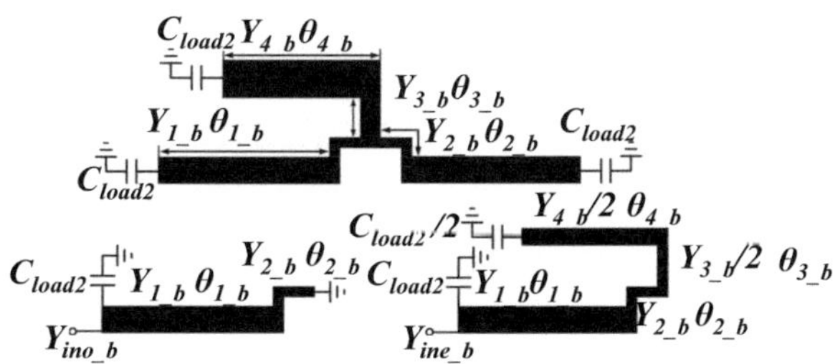

Fig. 4.12 Layout of Filter II-b and its even–odd-mode excitation equivalent circuit

To sum up, to achieve an independently tunable dual-band filter with constant absolute bandwidth of both passbands, three factors ideally need to be considered. Firstly, the independence of passbands tuning need to designed proportionally. Secondly, bandwidth of odd- and even-mode responses generated the passband must keep constant during the tuning range. Finally, the external quality factor Q_{ext} of passbands must satisfy the formulas presented in (4.43) during the entire tuning range.

$$Q_{ext\_p1} = \frac{g_0 g_1}{FBW_1} \qquad Q_{ext\_p2} = \frac{g_0 g_1}{FBW_2} \tag{4.43}$$

4.3 Validation of the Proposed Dual-Band Tunable BPF

In this section, three kinds of filter are designed, fabricated and measured based on the theoretical analysis in Sect. 4.2. The first one is based on creating independent modes for dual-band tuning, as described in Sect. 4.2.1. The second and third ones are based on creating independent signal paths for dual-band tuning.

4.3.1 Based on Creating Independent Modes for Dual-Band Tuning

A two-pole dual-band filter with the tunable first-passband fractional bandwidth and constant second-passband absolute bandwidth is designed as shown in Fig. 4.13. This filter structure achieves dual-band tunability through the establishment of independent resonant modes. It can achieve the following specifications:

First tuning passband: 0.77–1.0 GHz
First-passband fractional bandwidth: 21.0–25.0%
Second tuning passband: 1.57–2.0 GHz
Second-passband absolute bandwidth: 120 MHz
Number of poles: two
Type: Butterworth frequency response.

To derive the required design parameters for the tunable filter, a filter centering at $f_1 = 0.77$ GHz and $f_2 = 1.57$ GHz with 3-dB fractional bandwidth of 21.0% and 3-dB absolute bandwidth of 120 MHz, respectively, has been designed. The desired parameters are derived from the second-order low-pass prototype, which are given as $K_1 = 0.132$, $K_2 = 0.10$, $Q_{exo} = 8.5$, and $Q_{exe} = 19.2$. With these known design parameters, the frequency-dependent coupling-element values for the ideal tunable filter are plotted in Figs. 4.14 and 4.15 using dot line. The loaded capacitance C_{odd} and C_{even} are chosen to be 1.3 pF and 2.6 pF, respectively. The total length L and

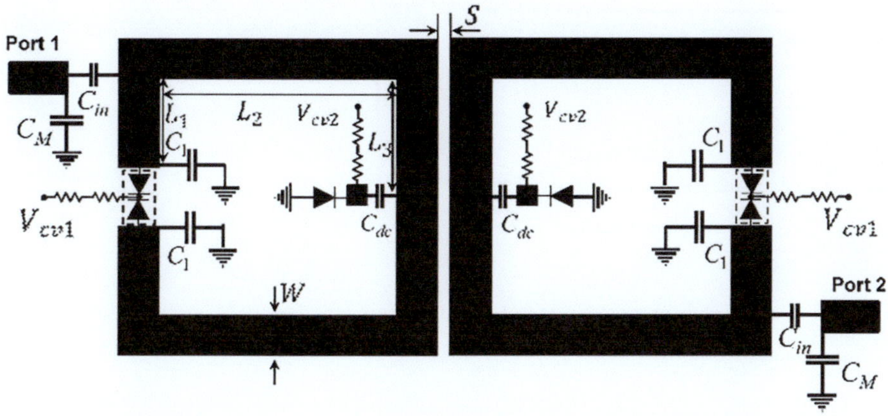

Fig. 4.13 Schematic of the proposed dual-pole dual-band tunable filter

capacitance C_1 can then be determined with respect to two central frequencies of the dual passbands, i.e., $f_1 = 0.77$ GHz and $f_2 = 1.57$ GHz. To achieve the desired values of K_1 and K_2 as indicated in Fig. 4.14a, b, respectively, the coupled-line parameters, L_3 and S, are chosen so that the simulated coupling coefficient could satisfy the desired values over the tuning range. The resultant parameters K_1 and K_2 are also plotted (solid line) in Fig. 4.14a, b. The parameters for the input matching capacitance C_{in} and C_M are chosen so that the extracted external quality to meet the requirement in the specified tuning range. The extracted Q_{exo} and Q_{exe} are also displayed (solid-line) in Fig. 4.15a, b, respectively.

To realize the above designed filter topology and desired frequency response, the dual-band tunable filter is designed and fabricated on a 0.8 mm F4B-2 substrate with $\varepsilon_r = 2.65$ using a single-layer copper etching PCB process. The proposed filter employs the varactor diodes with two types of packaging from the Skyworks: SMV1405-079LF for tuning C_{even}, and SMV1405-074LF for tuning C_{odd}, respectively. The single varactor capacitance is 0.63 pF and 2.67 pF at 30 V and 0 V reverse

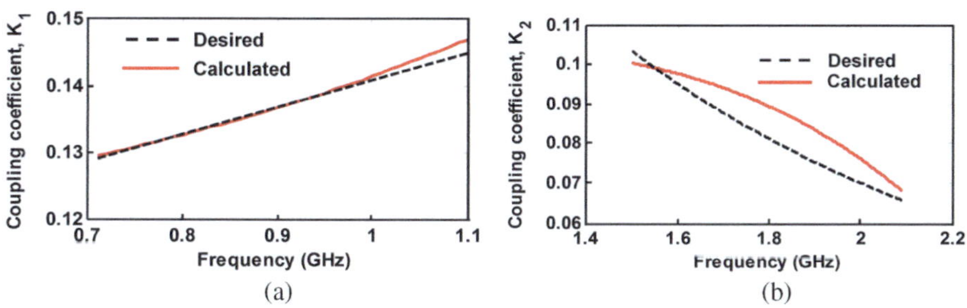

Fig. 4.14 Desired and simulated coupling coefficients of two coupled resonators, K_1 and K_2, in the dual passbands. **a** First passband. **b** Second passband

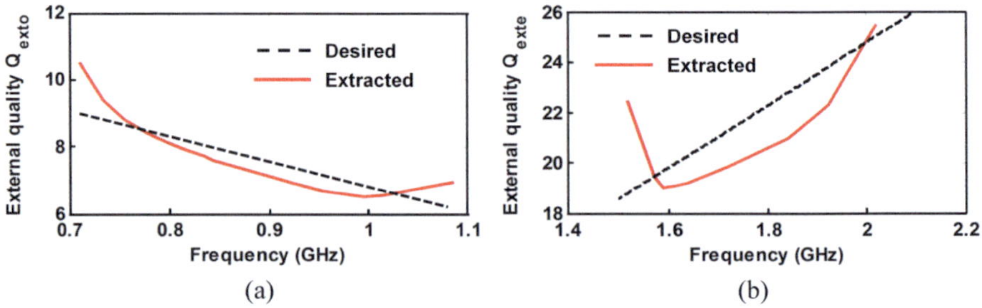

Fig. 4.15 Desired and extracted external quality factors for feeding a resonator in the dual passbands. **a** First passband. **b** Second passband

bias, respectively. The dc biasing circuit is formed by using two 100 kΩ resistances connected to reduce the RF-signal leakage through the bias network. The input/output lines are tapped at the resonator. Two capacitors chip C_{in} and two capacitors chip C_M are attached in the feeding lines, functioning as the input/output matching networks. Another two dc block capacitors C_{dc} are attached at the center of line. The parameters for the final layout of the proposed tunable filter, as referring to Fig. 4.13, are given in Table 4.1. The fabricated two-port filter is finally measured using the Agilent E5071C network analyzer.

Figure 4.16a, b show the simulated and measured results of the tunable dual-band filter with the fixed second passband and tunable first passband in their center frequencies. The 3-dB bandwidth and insertion loss are shown in Fig. 4.18a. The second passband center frequency is fixed at 1.70 GHz, and the first-passband center frequency is tuned from 0.77 to 1.0 GHz with the 3 dB fractional-bandwidth of 20.3–24.7%. Our results in Fig. 4.16 show that the measured return loss is better than 12 dB over the entire tuning range of the first passband while the insertion loss varies from 0.7 to 1.4 dB.

Figure 4.17a, b show the simulated and measured results of the tunable filter with the fixed first passband and tunable second passband center frequencies. The 3 dB bandwidth and insertion loss are shown in Fig. 4.18b. The first-passband center frequency is fixed at 1.0 GHz, and the second-passband center frequency is tuned from 1.57 to 2.0 GHz with the return loss better than 11 dB over the whole tuning range. Especially, the absolute bandwidth of the filter is kept as an approximate value of about 120 ± 8 MHz across the tuning range from 1.57 to 2.0 GHz with the insertion loss of 2.74–3.93 dB, which could be considered as a constant absolute bandwidth. From these results, it is clear that the dual-passband center frequencies of the proposed tunable filter can be tuned independently without any restriction in

Table 4.1 Parameters of proposed tunable dual-mode dual-band filter (unit: mm) (see Fig. 4.13)

W	$L_1/L_2/L_3$	S	$C_1/C_{in}/C_M/C_{dc}$ (pF)
1.0	9.0/10.0/9.5	0.2	0.5/15.0/1.6/15.0

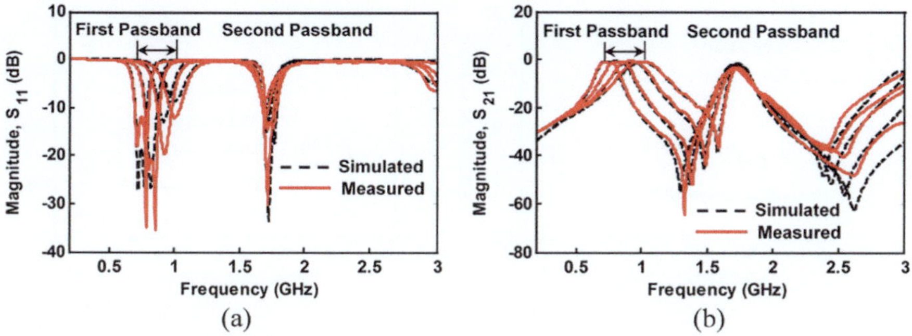

Fig. 4.16 Simulated and measured results of the tunable dual-band filter with the fixed second passband and adjustable first passband. **a** S_{11}-magnitude. **b** S_{21}-magnitude. Reverse-bias voltage variation: $V_{cv1} = 0 \sim 20$ V and $V_{cv2} = 1.8$ V

design and implementation. But, there exist some visible discrepancies in insertion loss between simulated and measured results at high frequencies and they are mainly caused by the approximate SPICE Model of the used varactor diodes and other unwanted parasitic effects including parasitic capacitance caused by wielding and the bias circuit in practical circuits. Figure 4.19 depicts the measured frequency responses under different input power levels to examine the power-handling capability. The results in Fig. 4.19 demonstrate that the presented tunable filter can handle the power in a range from 5 to 9 dBm without receiving any significant distortion in terms of the passband performance. The input third-order intercept point (IIP3) is measured by the Rohde and Schwarz FS300 spectrum analyzer and Agilent E4433B ESG-D series digital RF signal generator with 1 MHz frequency spacing. The measured IIP3 is found around 8.5 and 18.7 dBm at 0.77 and 1.0 GHz as well as 9.2 and 20.8 dBm at 1.57 and 2.0 GHz, respectively.

The photography of a fabricated filter is illustrated in Fig. 4.20 for the readers to see the configuration of a resultant tunable filter. The proposed two-order tunable

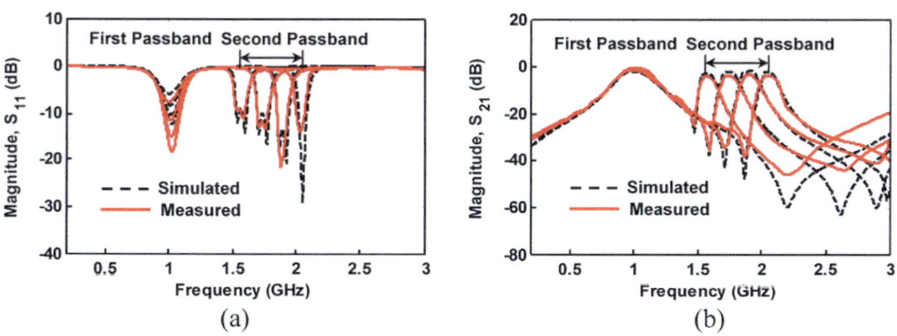

Fig. 4.17 Simulated and measured results of the tunable dual-band filter with the fixed first passband and adjustable second passband. **a** S_{11}-magnitude. **b** S_{21}-magnitude. Reverse-bias voltage variation: $V_{cv1} = 20$ V and $V_{cv2} = 0.5$–12 V

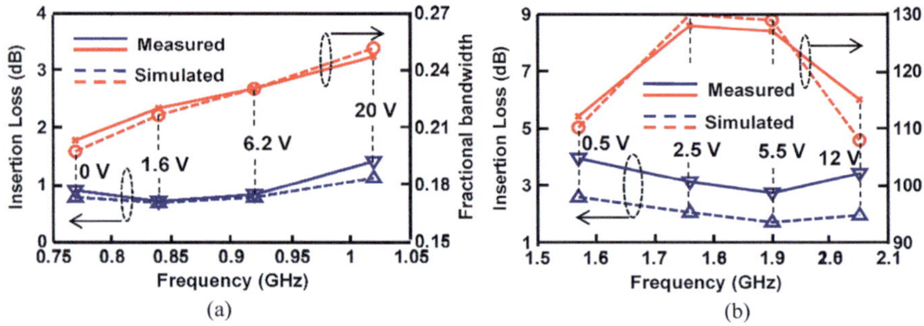

Fig. 4.18 Simulated and measured insertion loss and bandwidth of the dual-band tunable filter. **a** First passband. **b** Second passband

Fig. 4.19 Measured S_{21} with different input powers at bias voltages: $V_{cv1} = 10$ V and $V_{cv2} = 3$ V

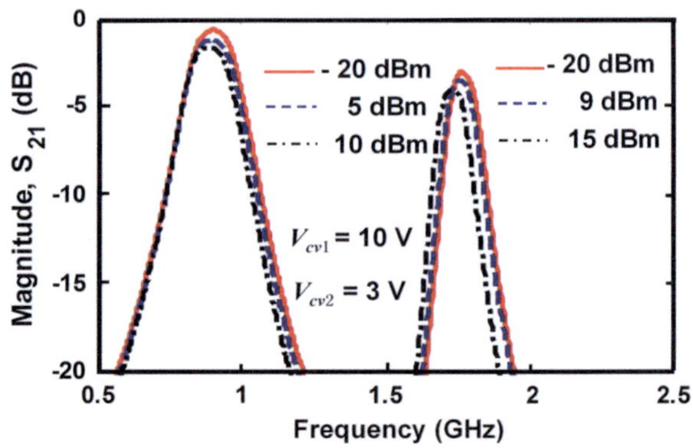

filter can provide independent dual passbands characteristics with a simple coupling topology and less types of control voltages.

4.3.2 Based on Independent Signal Paths for Dual-Band Tuning

Filter I

By combining Filter I-a and Filter I-b in the above theoretical analysis, this Filter I achieves dual-band tunability by establishing independent signal transmission paths. The external quality factors, Q_{e1} and Q_{e2} of the two bands can be extracted [13], respectively. As shown in Fig. 4.21a, when the coupled gap width S_1 varies from 0.1 to 0.4 mm, Q_{e1} increases from 28.9 to 122. Simultaneously, little change of Q_{e2} basically stays the same. As shown in Fig. 4.21b, when S_2 changes from 0.1 to 0.4 mm, Q_{e2} increases from 11.2 to 17.6, and Q_{e1} just remains in a narrow band 28.9–25.0. Additionally, a detailed analysis of the coupling gap g between the input

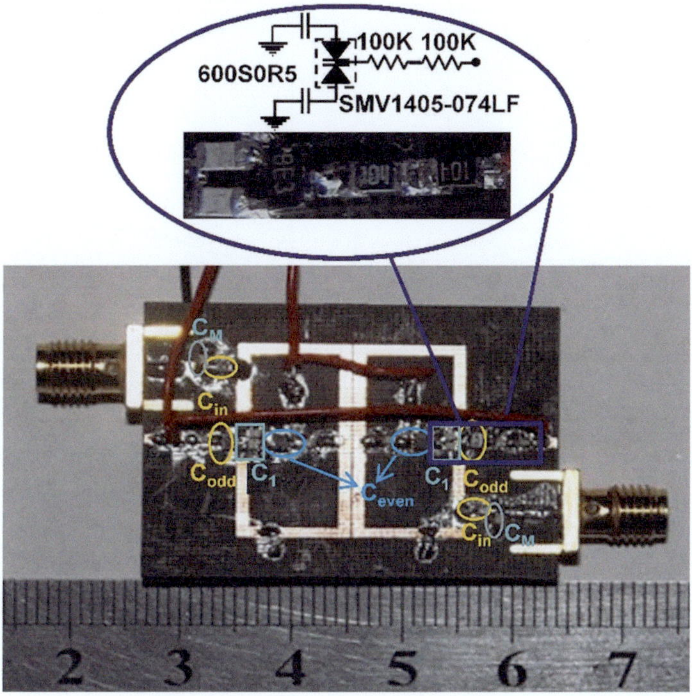

Fig. 4.20 Photograph of the fabricated tunable dual-band filter

and output ports is shown in Fig. 4.21c. The gap g has litter impact on the insertion loss and the coupling. By observing Fig. 4.21, the design guidelines of the feed circuit can be provided and the initial feed part of the presented filter can be determined independently.

From the analysis above, we can achieve a configurable BPF with a predefined constant fractional-bandwidth first passband and a controllable second passband with constant absolute bandwidth. The simulation results are shown in Fig. 4.22. Obviously, the first passband shows a tuning range 1.449–1.889 GHz, insertion loss of 3.642–1.967 dB, a constant fractional bandwidth of 4.35 ± 0.55%, and the second passband remains constant during the tuning process. The second passband can be controlled during the frequency range of 2.26–3.26 GHz with an absolute bandwidth of 10.5 ± 11 MHz, an insertion loss of 2.265–3.165 dB, as shown in Fig. 4.22b. Meanwhile, the first passband can hardly be influenced.

The design procedure of the presented configurable dual-band BPF is given as follows. Firstly, determine the dimensions of Filter I-a and Filter I-b to achieve the tuning ranges and bandwidths; Then determine the feed line dimensions according to Fig. 4.21; Finally, the optimized dual-band configurable BPF can be achieved.

To demonstrate the performance of the proposed filter, a tunable filter is designed and fabricated, as shown in Fig. 4.23. The dimensions for the filter are presented in Table 4.2 for $\varepsilon_r = 2.65$, $h = 0.5$ mm with $w_4 = 2$ mm, $s_1 = 0.1$ mm, $s_2 = 0.2$ mm.

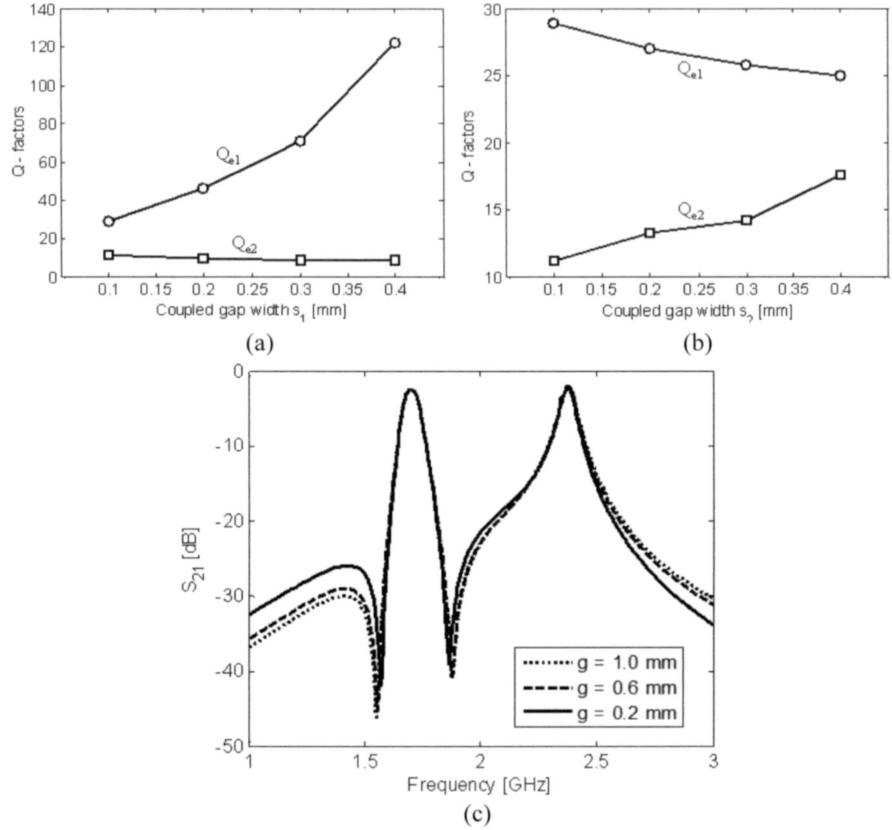

Fig. 4.21 Extracted and simulated results of Filter I. **a** Q-factors as function of s_1 ($s_2 = 0.1$ mm). **b** Q-factors as function of s_2 ($s_1 = 0.1$ mm). **c** S_{21} as function of g ($s_1 = 0.1$ mm, $s_2 = 0.2$ mm)

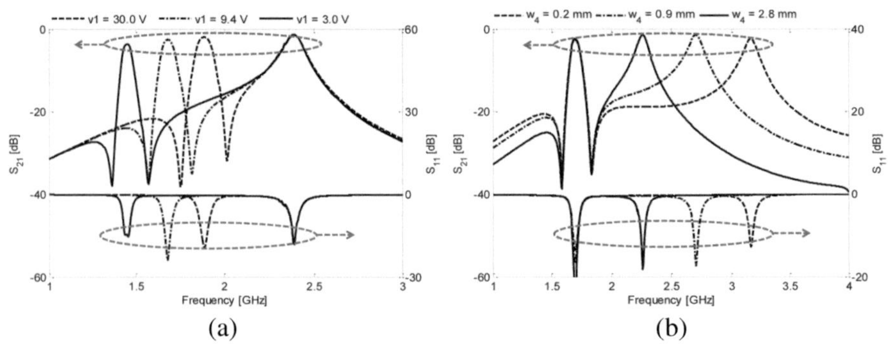

Fig. 4.22 Tuning range of passbands. **a** First passband under different bias voltage ($s_1 = 0.1$ mm, $s_2 = 0.1$ mm). **b** Second passband under different width w_4 ($s_1 = 0.1$ mm, $s_2 = 0.1$ mm)

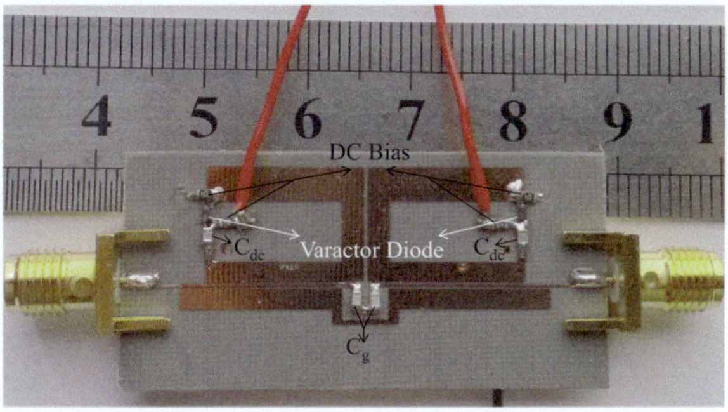

Fig. 4.23 Photograph of fabricated configurable dual-band BPF

The dc bias is done by using a 100 kΩ resistor and an ATC chip capacitor between the DC voltage and the open ends of the resonator. $v1$ is the bias voltage of C_{L1}. The capacitor-varactor series connection hardly changed the overall capacitance with the chip capacitor value 15 pF. The C_g is selected as 0.6 pF as a tradeoff between tunability and loss.

The simulated and measured results are presented in Fig. 4.24. Clearly, the first passband shows a constant fractional-bandwidth range of 3.9 ± 0.3%, a center frequency of 1.472–1.886 GHz, insertion loss of 4.2–5.7 dB while the bias voltage $v1$ changes from 30.0 to 3.0 V. Among the first passband tuning process, the second passband exhibits a small variation with center frequency of 2.24 GHZ, an absolute bandwidth of 53 MHz, an insertion loss of 3.517 dB. The simulation and measurement show good agreement. The distinction between the simulation and the measurement is mainly due to the mis-machining tolerance.

Filter II

By combining Filter II-a and Filter II-b in the above theoretical analysis, a novel independent tunable dual-bandpass filter with constant absolute bandwidth is constructed, as shown in Fig. 4.25. Since the two dual-mode resonators are only coupled to the IN/OUT ports, two independent passbands can be generated. With only two bias voltages, the two passbands can be simply tuned, and the absolute bandwidth of 3 dB remains constant over the tuning range. Here, two chip capacitors, C_g, are used as matching capacitors.

Based on the proposed filter structure, there are four types of dual-band selectivity to choose while designing the tunable filter, as depicted in Fig. 4.28. The first even- and odd-mode of resonator Filter II-a, f_{e_a} and f_{o_a} form the first passband. Then the second passband are generated by the first even- and odd-mode of resonator Filter

Table 4.2 Critical dimensions of the bandwidth tunable BPF with harmonic suppression (unit. mm)

l_1	l_2	l_3	l_4	l_5	w_1	w_2	w_3
11.8	5.6	2.2	15.1	9.0	0.8	3.3	2.0

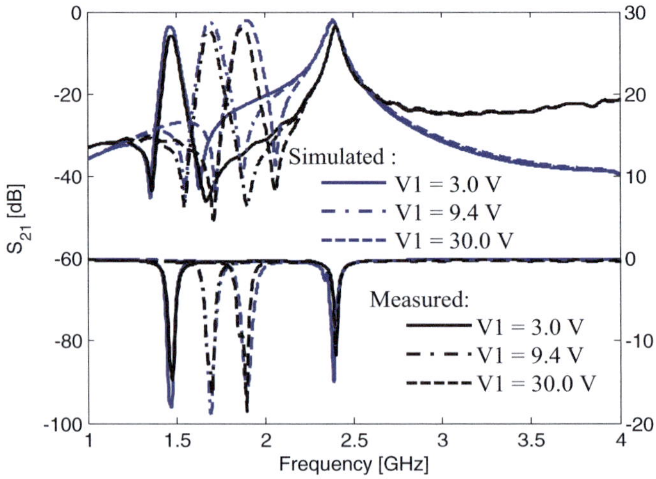

Fig. 4.24 Simulation and measurement of fabricated configurable dual-band BPF

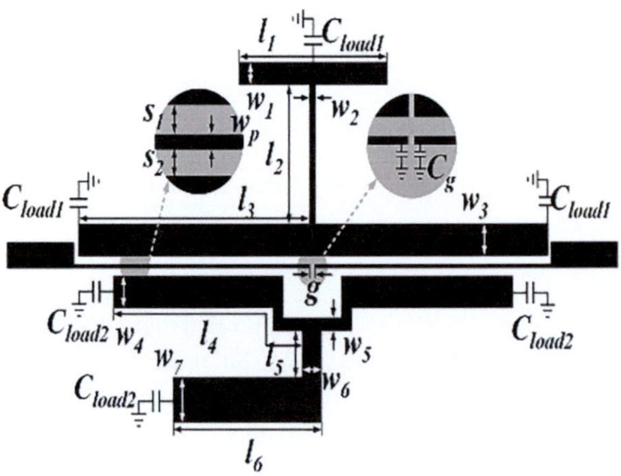

Fig. 4.25 Layout of the proposed dual-band tunable filter

II-bf_{e_b} andf_{o_b}. Three transmission zeros are introduced at the adjacent of the passbands to improve the selectivity. Among them, the first and third transmission zeros are mainly generated by the dual-mode resonators, respectively [14]. The second transmission zero is mainly due to the multipath effect [17].

After that, in order to confirm the constant odd- and even-mode bandwidth, parameters of admittance Y and electrical length θ should be chosen properly. To demonstrate how to achieve this, the dual-band selectivity shown in Fig. 4.27a are selected. The parameters are given as follows: $Y_{1\_a} = 0.025$ S, $l_{1\_a} = 17.8$ mm, $Y_{2\_a} = 0.011$ S, $l_{2\_a} = 8.4$ mm, $Y_{3\_a} = 0.025$ S, $l_{3\_a} = 6.4$ mm, $Y_{1\_b} = 0.025$ S, $l_{1\_b} = 12.8$ mm, $Y_{2\_b} = 0.0167$ S, $l_{2\_b} = 2.6$ mm, $Y_{3\_b} = 0.0182$ S, $l_{3\_b} = 5.5$ mm, $Y_{4\_b} = 0.278$ S, $l_{4\_b} = 11.7$ mm. $l_{n\_a,b}$ are the corresponding physical length of $\theta_{n\_a,b}$ (n = 1, 2, 3).

To reduce the complexity of choosing appropriate parameters, here is a method to confirm the parameters quickly. Firstly, the odd-modes could be calculated according to Eqs. (4.37) and (4.40) and the type of selectivity presented in Fig. 4.26. As $|f_o - f_e|$ is constant, the desired f_e can be confirmed. Then, adjust the even mode parts which are independent of odd mode parts to approach the curve of the desired even-modes. The results are shown in Fig. 4.27a. It is notable that the $|f_o - f_e|$ of resonator Filter II-a and Filter II-b can almost keep constant between the variable loading capacitance in 0.3–1.3 pF. Finally, to satisfy the requirement of Q_{ext}, the matching part, s_1, s_2, w_p, C_g should be designed properly. The desired and extracted Q_{ext} are presented at Fig. 4.27b. The extracted results show that the Q_{ext} can basically meet the design requirements.

To demonstrate this type of filter experimentally, the filter was fabricated on a 0.508-mm Rogers RT/Duroid 5880 substrate ($\varepsilon_r = 2.2$, $\tan\delta = 0.0009$) using a copper etching process. The photograph of the fabricated filter is presented in Fig. 4.28. The dimensions are as follows: $l_1 = 6.4$, $l_2 = 8.4$, $l_3 = 17.5$, $l_4 = 12.4$, $l_5 = 2.8$, $l_6 = 11.7$, $w_1 = 1.4$, $w_2 = 0.6$, $w_3 = 2.2$, $w_4 = 2.0$, $w_5 = 1.2$, $w_6 = 1.6$, $w_7 = 2.4$, $s_1 = 0.2$, $s_2 = 0.2$, $g = 0.4$, $w_p = 0.4$ (Unit: mm). The terminal loaded capacitors C_{load} were realized by connecting a SKYWORKS SMV-1405 with SC-70 package. The single varactor capacitance is 0.63 pF and 2.67 pF at 30 V and 0 V reverse bias, respectively. The DC biasing is done using a 100 K resistor. The fixed capacitor C_g value of 0.6 pF was chosen to meet the requirement of Q_{ext}. The S-parameters of the filter were measured with an Agilent E5071C vector network analyzer.

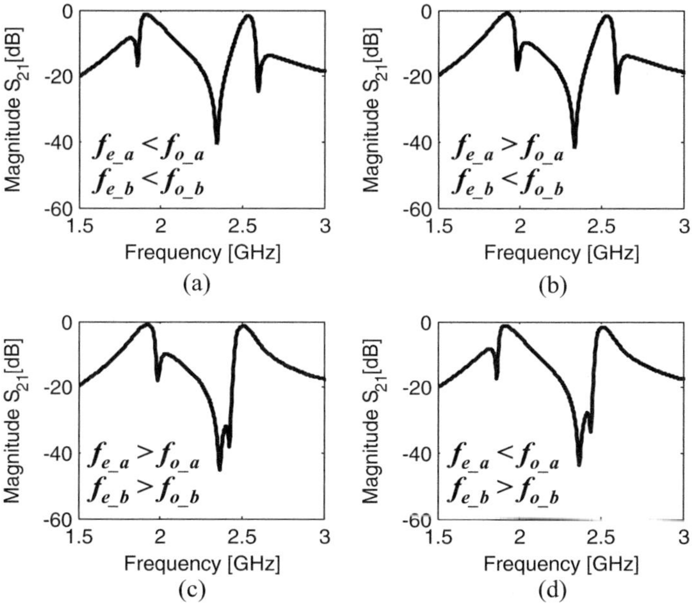

Fig. 4.26 Dual-band selectivity under different conditions of f_{e_a}, f_{o_a}, f_{e_b} and f_{o_b}

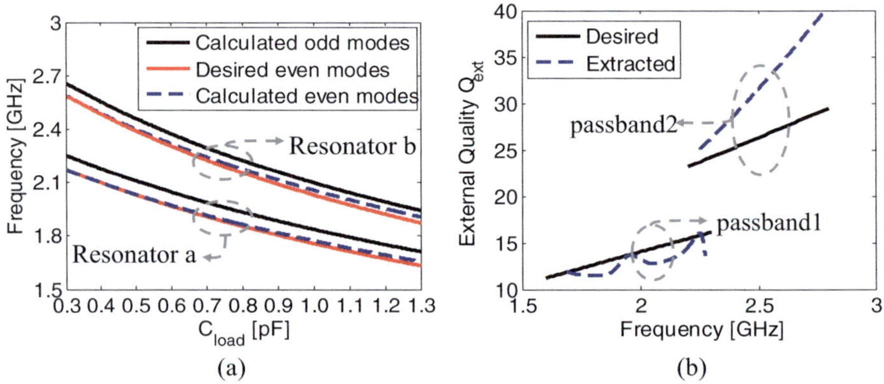

Fig. 4.27 Simulated results. **a** Resonant modes values varied with the C_{load} under different capacitance. **b** Desired and extracted external quality of passbands

Fig. 4.28 Photograph of the fabricated filter

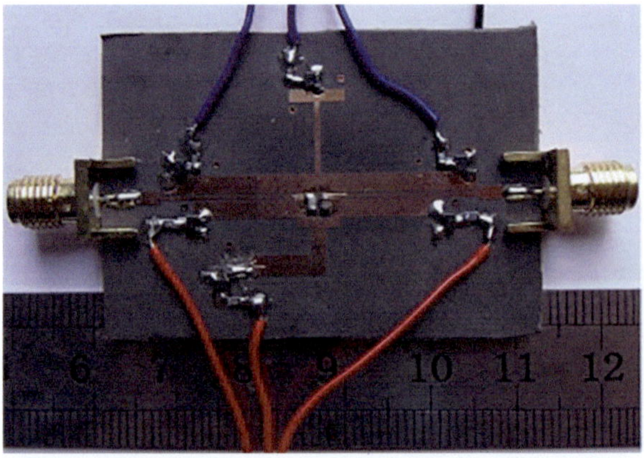

The measured S_{21} and S_{11} are plotted in Fig. 4.29a, b, respectively. As Fig. 4.29a shown, the second passband center frequency is fixed at 2.625 GHz, and the first passband center frequency is tuned from 1.696 to 2.272 GHz with the return loss better than 12 dB over the entire tuning range. Especially, the absolute bandwidth of the filter is about 130 ± 10 MHz across the tuning range from 1.70 to 2.27 GHz with the insertion loss of 3.65–4.55 dB, which could be considered constant absolute bandwidth. Then as depicted in Fig. 4.29b, the first passband center frequency is fixed at 1.686 GHz, and the second passband center frequency is tuned from 2.248 to 2.77 GHz with the return loss better than 10 dB over the entire tuning range. Especially, the absolute bandwidth of the filter is about 80 ± 10 MHz across the tuning range from 2.25 to 2.77 GHz with the insertion loss of 4.11–4.87 dB, which could also be considered constant absolute bandwidth.

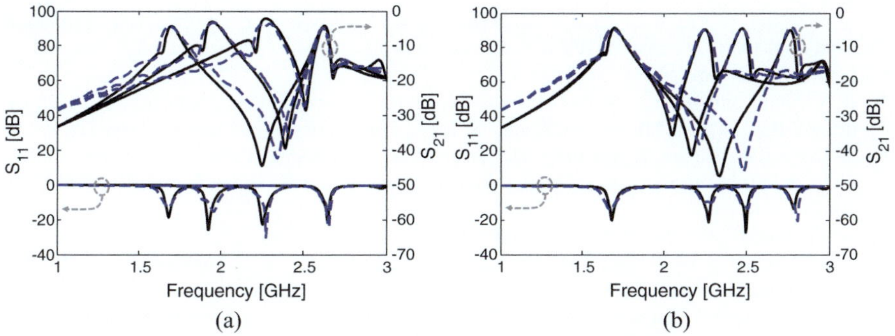

Fig. 4.29 Simulation and measured results of the dual-band tunable filter. Black solid lines and blue dotted lines stand for the simulation and measured results, respectively. **a** S-parameters with fixed first passband and tunable second passband. **b** S-parameters with fixed first passband and tunable second passband

4.4 Conclusion

This chapter introduces two methodologies for realizing dual-band tunable bandpass filters, presenting multiple filter prototypes developed under these frameworks. The multi-mode excitation approach achieves functionality by independently exciting and controlling distinct resonant modes within a single resonator. An example in Sect. 4.3.1 demonstrates a filter based on a tunable dual-mode resonator, where varactor diodes loaded at specified locations independently regulate even/odd modes, enabling decoupled center frequency tuning for both passbands. Conversely, the independent signal path architecture centres on constructing mutually isolated transmission channels. Two examples in Sect. 4.3.1 use two non-interfering coupling schemes to generate dual-band responses, exhibiting strong alignment between simulation and measurement data. Collectively, these implementations confirm that either mode-based or path-isolated control strategies-when coupled with optimized structural design-can achieve compact dual-band filters with fully independent passband tunability, thus advancing RF system design capabilities.

References

1. X.-H Guan, W. Fu, G.-H Li, S. Jiang and H.-W Liu, Novel microstrip dual-band bandpass filter with wide stopband and high isolation, Microwave Optical Technology Letter, vol. 53, pp. 803–806, 2011.
2. C.-W Tang and P.-H Wu, Design of a planar dual-band bandpass filter, IEEE Microwave Wireless Component Letter, vol. 21, pp. 362–364, 2011.
3. L. Gao, X.-Y Zhang, Novel dual-band bandpass filters using stub-loaded short-ended resonators, Microwave Optical Technology Letter, vol. 54, pp. 2771–2774, 2012.

4. X.-S Zhang, B. Liu, Y.-J. Zhao, J.-K. Wang and W. Chen, Compact and high selectivity dual-band dual-mode microstrip BPF with five transmission zeros, Microwave Optical Technology Letter, vol. 54, pp. 79–81, 2012.
5. X. Y. Zhang and Q. Xue, "Novel centrally loaded resonators and their applications to bandpass filters," IEEE Trans. Microwave Theory & Tech., vol. 56, no. 4, pp. 913–921, Apr. 2008.
6. D. Girbau, A. Lazaro, E. Martinez, D. Masone, and, L. Pardell, "Tunable dual-band bandpass filter for WLAN applications," Microw. Opt. Technol. Lett., vol. 51, no. 9, pp. 2025–2028, Sep, 2009.
7. G. Chaudhary, H. Choi, Y. Jeong, J. Kim, D. Kim and J.C. Kim, "Design of dual-band bandpass filter using DGS with controllable Second passband," IEEE Microwave and Wireless Components Letters, vol. 21, pp. 589–591, 2011.
8. E. E. Djoumessi, M. Chaker, and K. Wu, "Varactor-tuned quarter-wavelength dual-bandpass filter," IET Microw. Antenna and Propagat., vol. 3, no. 1, pp. 117–124, 2009.
9. G. Chaudhary, Y. Jeong, and J. Lim, "Harmonic suppressed dual-band bandpass filter with tunable passbands," IEEE Trans. Microwave Theory Tech., vol. 60, no. 7, pp. 2115–2123, Jul, 2012.
10. G. Chaudhary, Y. Jeong, and J. Lim, "Dual-band bandpass filter with independently tunable center frequencies and bandwidths," IEEE Trans. Microwave Theory Tech., vol. 61, no. 1, pp. 107–116, Jan, 2013.
11. X. Huang, L. Zhu, Q. Feng, Q. Xiang and D. Jia, "Tunable Bandpass Filter With Independently Controllable Dual Passbands," in IEEE Transactions on Microwave Theory and Techniques, vol. 61, no. 9, pp. 3200–3208, Sept. 2013, https://doi.org/10.1109/TMTT.2013.2273894.
12. S. J. Park, M. A. El-Tanani, I. Reines, and G. M. Rebeiz, "Low-loss 4–6-GHz tunable filter with 3-bit high-Q orthogonal bias RF-MEMS capacitance network," IEEE Trans. Microw. Theory Tech., vol. 56, no. 10, pp. 2348–2355, Oct. 2008.
13. J. S. Hong and M. J. Lancaster, Microstrip filter for RF/Microwave Applications. New York: Wiley, 2001.
14. W. X. Tang and J. S. Hong, "Varactor-tuned dual-mode bandpass filters," IEEE Trans. Microw. Theory Tech., vol. 58, no. 8, pp. 2213–2219, Aug. 2010.
15. Ding-Hong Jia, Quanyuan Feng, Xiao-Guo Huang, and Qian-Yin Xiang, "A Dual-Band Bandpass Filter with a Tunable Passband," Progress In Electromagnetics Research C, Vol. 37, 107–118, 2013. https://doi.org/10.2528/PIERC12121210.
16. D. Jia, Q. Feng, X. Huang and Q. Xiang, "A constant absolute bandwidth filter with dual independent tunable passbands," 2014 IEEE International Wireless Symposium (IWS 2014), Xi'an, China, 2014, pp. 1–4, https://doi.org/10.1109/IEEE-IWS.2014.6864192.
17. L. Li and D. Uttamchandani, "Demonstration of a tunable RF MEMS bandpass filter using silicon foundry process," Journal of Electromagnetic Waves and Applications, vol. 23, pp. 405–413, 2009.

Chapter 5
Tunable Bandstop Filters

5.1 Introduction

In previous chapters, we analyzed the design theory and validation of tunable bandpass filters. In this chapter, configurable and tunable bandstop filters are presented. The design theory and validation of configurable and tunable bandstop filters are presented.

Due to their effective suppression of spurious signals in wireless communication systems, bandstop filters (BSFs) are highly desired. Among these papers [1–6], some effective design methods have been presented. Stepped-impedance resonator (SIR) and spur-lines were applied to microstrip structures to widen the bandwidth of stopband response in [1]. Cross-coupled lines were introduced to design quasi-elliptic BSF in [2]. In [4, 5], different microstrip structures were adopted to realize wideband BSFs. Capacitors were connected to microstrip structure to design BSFs in [6]. To the best of the author's knowledge, the BSFs with independently controllable bandwidth and center frequency have the higher adaptability in practical applications, which allows to independently control the bandwidth and the center frequency. However, the independence between the bandwidth and the center frequency was not involved in [1–6].

Meanwhile, dual-band bandstop filters (DBBSFs) are also attractive owing to their ability to treat the double-sideband spectral regrowth of the output signals of high power amplifiers by using one filter and to reduce circuit size and cost. Some valid design methods of dual-band bandstop filters (DBBSFs) have been presented in the published papers [7–15]. A two-step frequency-variable transformation was applied to the low-pass prototype to produce dual-stopband responses in [7]. In [8], metamaterial transmission lines were used to design DBBSFs. Coupled stepped-impedance-resonators were used to design DBBSF in [9]. Open stubs were connected to the main transmission line to achieve bandstop response in [10–12]. The center frequency and bandwidth can be controlled by the impedance ratio and the lengths of the stubs. Defected ground structures (DGS) were also utilized to achieve bandstop responses

F. Nian et al., *RF & Microwave Multi-mode and Tunabe Filters*,
https://doi.org/10.1007/978-981-92-0577-6_5

in [13, 14]. In order to prevent performance degradation, which is caused by the transmission lines between resonators, Cameron et al. adopted synthesis technique to design DBBSFs with direct-coupled configuration [15]. However, the independence between the two stopbands in terms of center frequency and bandwidth is hardly involved in most of these papers. In addition, further size compactness and high rejection levels at mid-stopbands are still needed.

Tunable bandstop filters (BSFs) are useful to prevent unwanted signals from interfering wireless communications and relax the requirement for bandpass filter selectivity [16–20]. Furthermore, as the channel bandwidth is usually fixed no matter how the carrier frequency changes, tunable bandstop designs with a constant absolute bandwidth (CABW) are desirable within the tuning range for many applications regardless of the tuning technologies. Some works on this issue have been reported in the literature. In [18], the bandstop filter employed varactor-tuned resonator pairs. Separate bias was applied to each of the two varactors in the resonator pair to obtain the desired center frequencies and keep the bandwidth constant. The filter in [19] was designed based on a dual-band structure consisting of a wideband bandpass filter with an integrated bandstop filter. When the resonator has an electric length of 52°, the absolute stopband bandwidth remains unchanged. In [20], a doublet configuration was adopted to design tunable bandstop filters. Mixed electronic and magnetic coupling scheme was introduced to keep the bandwidth constant. The authors provided an effective method to design narrow-band tunable bandstop filters with constant bandwidth.

In this chapter, firstly, novel bandstop units with independently controlled bandwidth and center frequency are presented. Derived from the theoretical analysis, the center frequencies and bandwidth of the single- and dual-stopband units can both be independently controlled, which highly improves the design freedom. Based on the single and dual stopband units, compact BSFs and DBBSF are designed, fabricated and measured. The measured performance of the fabricated filter shows good agreement with the proposed theory. Then, based on the proposed bandstop unit and its controllable bandstop characteristics, a novel approach of designing tunable wideband bandstop filters with CABW is proposed. A simple bandstop unit is adopted to design tunable filters. The bandstop unit shows a unique property that its frequency can be easily tuned and keep the absolute bandwidth (ABW) constant over the tuning range. Compact J-inverters are used to cascade the bandstop units. The tunable BSF design shows a wide tuning range and high rejection levels at mid-stopband. For demonstration, tunable bandstop filters are fabricated and measured.

Based on this, the subsequent sections of this chapter will proceed as follows: Sect. 5.3 will provide an in-depth analysis of the operating principles of the proposed bandstop filters. Section 5.4 will introduce the validation for the proposed bandstop filters, present their simulated and measured results. Finally, Sect. 5.5 will present a comprehensive summary of the chapter's content.

5.2 Bandstop Unit Design

In this section, the design methods of the sing-stopband/dual-stopband/tunable stopband resonant units are proposed. Their characteristics of stopband are theoretical analyzed and simulated.

5.2.1 Single-Stopband Resonant Unit

As illustrated in Fig. 5.1, the proposed single-bandstop unit consists of two transmission lines (Y_{t1}, θ_{t1}) and a center loaded stepped-impedance stub (Y_{t2}, θ_{t2} and Y_{t3}, θ_{t3}). As the single-bandstop unit has a simple structure, the design process is significantly simplified [21]. Afterwards, the characteristic impedances of the microstrip lines at the input/output ports are set as 50 Ω.

Thus, the overall *ABCD* matrix of the single-bandstop unit can be deduced as

$$\begin{bmatrix} A_u & B_u \\ C_u & D_u \end{bmatrix} = A_{t1} \times A_{stub} \times A_{t1} \tag{5.1}$$

where

$$A_{t1} = \begin{bmatrix} \cos\theta_{t1} & j\sin\theta_{t1}/Y_{t1} \\ jY_{t1}\sin\theta_{t1} & \cos\theta_{t1} \end{bmatrix}, \quad A_{stub} = \begin{bmatrix} 1 & 0 \\ Y_{sub} & 1 \end{bmatrix},$$

$$A_u = D_u = \cos 2\theta_{t1} + jY_{sub}\sin 2\theta_{t1}/2Y_{t1},$$

$$B_u = -Y_{sub}\sin^2\theta_{t1}/Y_{t1}^2 + j\sin 2\theta_{t1}/Y_{t1},$$

$$C_u = Y_{sub}\cos^2\theta_{t1} + jY_{t1}\sin 2\theta_{t1}, \quad Y_{sub} = jY_{t2}\cdot\frac{2Y_{t3}\tan\theta_{t3} + Y_{t2}\tan\theta_{t2}}{Y_{t2} - 2Y_{t3}\tan\theta_{t3}\tan\theta_{t2}}.$$

Then, the insertion loss of the single-bandstop unit can be achieved from *A* matrix by [22]

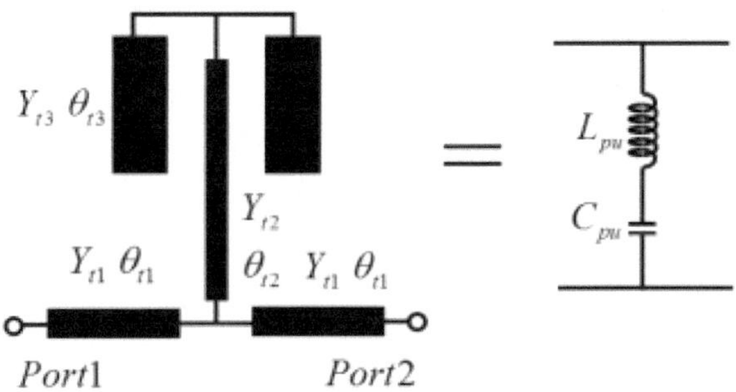

Fig. 5.1 Layout of the proposed single-bandstop unit and its equivalent circuit

$$S_{21} = \frac{2}{A_u + B_u/Z_0 + C_u Z_0 + D_u} \tag{5.2}$$

In order to suppress the desired frequency band, the insertion loss of the center frequency of the single-stopband must equal to zero, which yields

$$\frac{1}{R} = 2\tan\left(\frac{2\pi f\sqrt{\varepsilon_r}l_2}{c}\right)\tan\left(\frac{2\pi f\sqrt{\varepsilon_r}l_3}{c}\right) \tag{5.3}$$

where $R = Y_{t3}/Y_{t2}$, c is the light speed in the free space, ε_r is the dielectric constant of the substrate and l_1, l_2, l_3 stands for the corresponding physical length of θ_{t1}, θ_{t2}, θ_{t3}, respectively.

Equation (5.3) indicates that the center frequency of the single-bandstop unit is independent of the transmission lines (Y_{t1}, θ_{t1}). Furthermore, it is mainly determined by R, l_2, and l_3. It can be inferred that the center frequency can be freely tuned against the physical lengths of the stub while keeping R constant.

The equivalent circuit of the single-bandstop unit is also presented in Fig. 5.1. The stub can be equivalent to the series connected inductor L_{pu} and capacitor C_{pu}. The frequency mapping from low-pass response to the bandstop response yields the *LC* elements [22] as

$$L_{pu} = \frac{1}{\omega_{0u} g_{1u} \Delta_u}, \quad C_{pu} = \frac{g_{1u}\Delta_u}{\omega_{0u}} \tag{5.4}$$

where g_{1u} is the value of the low-pass net element, Δ_u is the fractional bandwidth (FBW) of the unit, ω_{0u} is the center angular frequency of the unit.

As discussed before, the center frequency of the single-stopband is mainly decided by the parameters of the loaded stub. Thus, the loaded stub and its equivalent circuit must exhibit the same reactance slope b to obtain the same bandwidth. Consequently, the reactance slope parameter b can be used to analyze the bandwidth [23]. It can be written as

$$x = \frac{\omega_0}{2} \cdot \frac{dX}{d\omega}\Big|_{\omega=\omega_0} \tag{5.5}$$

Then, the equation to calculate the bandwidth can be written as

$$\frac{\omega_{0u}}{2} \cdot \frac{dX_{load}}{d\omega}\Big|_{\omega=\omega_{0u}} = \frac{1}{g_1 \Delta_u} \tag{5.6}$$

where X_{load} can be achieved as

$$X_{load} = -\frac{1}{Y_{t2}} \cdot \frac{1 - 2R_1 \tan\theta_{t3}\tan\theta_{t2}}{(2R_1 \tan\theta_{t3} + \tan\theta_{t2})} \tag{5.7}$$

By the analysis above, Eqs. (5.3) and (5.6) clearly reveals the center frequency and bandwidth of the stopband, respectively. Then, a full-wave simulation is used to include all of parasitic effects such as open-end fringing, right-angle bend and via-hole inductance in the filter design. According to Eq. (5.3), the center frequency of the proposed unit can be freely tuned by varying l_2 or l_3 while keeping the impedance ratio R constant. As depicted in Fig. 5.2a, taking l_2 for example, the stopband move to the lower frequency as l_2 increases from 6.5 to 12.0 mm. Figure 5.2b shows that the variation of the bandwidth of the bandstop unit versus various impedance ratio R and Y_{t2}. As presented by Eqs. (5.5)–(5.7), the bandwidth can be enlarged by increase the values of Y_{t2}. Then, in order to keep the center frequency unmoved, R is tuned to a smaller value. At last, the bandwidth of the stopband can be enlarged without changing the center frequency. This method introduces the design freedom to the bandstop unit and can be adopted to design BSFs with independently controllable bandwidth and center frequency.

5.2.2 *Dual-Stopband Resonant Unit*

Figure 5.3 illustrates the configuration of the proposed dual-stopband unit and its equivalent circuit, which is transformed by the bandstop unit indicated in Fig. 5.1. It consists of two transmission lines (characteristic admittance $Y_{\mathrm{b}1}$, electrical length θ_{b1}) and two center loaded stepped-impedance stubs (loaded stub b1: characteristic admittance Y_{b2}, Y_{b3}, electrical length θ_{b2}, θ_{b3}; loaded stub b2: characteristic admittance Y_{b4}, Y_{b5}, electrical length θ_{b4}, θ_{b5}). Moreover, the characteristic impedance of the microstrip lines at the input/output ports is $Z_0 = 50\ \Omega$.

Then, the overall *ABCD* matrix of the dual-band bandstop unit can be written as

$$
\begin{aligned}
A = D &= \cos 2\theta_{b1} + j(Y_{loadb1} + Y_{loadb2}) \sin 2\theta_{b1}/2Y_{b1},\\
B &= -(Y_{loadb1} + Y_{loadb2}) \sin^2 \theta_{b1}/Y_{b1}^2 + j \sin 2\theta_{b1}/Y_{b1},\\
C &= (Y_{loadb1} + Y_{loadb2}) \cos^2 \theta_{b1} + jY_{b1} \sin 2\theta_{b1},
\end{aligned}
$$

where $R_{b1} = Y_{b3}/Y_{b2}$, $R_{b2} = Y_{b5}/Y_{b4}$,

$$
Y_{loadb1} = jY_{b2} \cdot \frac{2R_{b1} \tan\theta_{b3} + \tan\theta_{b2}}{1 - 2R_{b1} \tan\theta_{b3} \tan\theta_{b2}}, \quad Y_{loadb2} = jY_{b4} \cdot \frac{2R_{b2} \tan\theta_{b5} + \tan\theta_{b4}}{1 - 2R_{b2} \tan\theta_{b5} \tan\theta_{b4}}.
$$

The insertion loss of ABCD matrix can also be achieved by Eq. (5.2). To suppress the desired frequency bands, the insertion loss must be equal to zero at the center frequency of the stopband. By equating $|S_{21}| = 0$, it yields that $Y_{loadb1} \to \infty$, $Y_{loadb2} \to \infty$. Then, we obtain

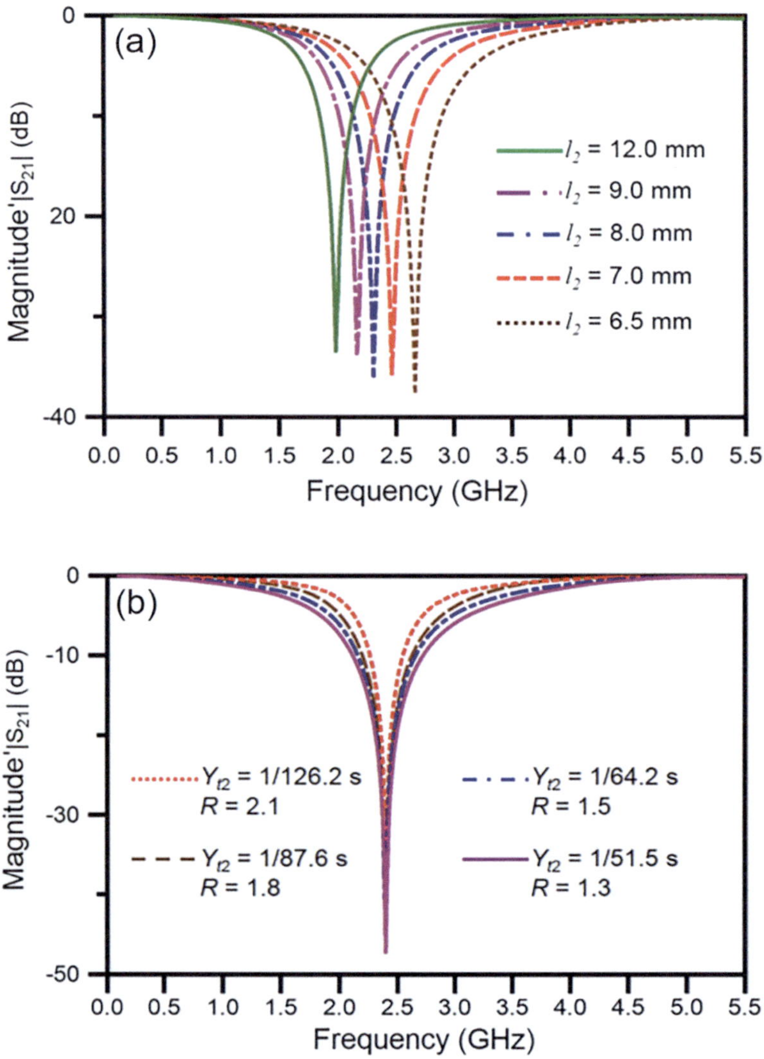

Fig. 5.2 Simulated results. **a** Tuning of center frequency against l_2 ($l_3 = 5.6$ mm, $R = 1.8$). **b** Bandwidths variation against admittance ratio R and Y_{t2} with constant $l_2 = 8.1$ mm and $l_3 = 5.6$ mm

$$
\begin{aligned}
\frac{1}{R_{b1}} &= 2\tan\left(\frac{2\pi f\sqrt{\varepsilon_r}l_{b2}}{c}\right)\tan\left(\frac{2\pi f\sqrt{\varepsilon_r}l_{b3}}{c}\right),\\
\frac{1}{R_{b2}} &= 2\tan\left(\frac{2\pi f\sqrt{\varepsilon_r}l_{4b}}{c}\right)\tan\left(\frac{2\pi f\sqrt{\varepsilon_r}l_{5b}}{c}\right)
\end{aligned}
\tag{5.8}
$$

where l_{bn} denotes the corresponding physical length of θ_{bn} ($n = 1, 2, 3, 4, 5$).

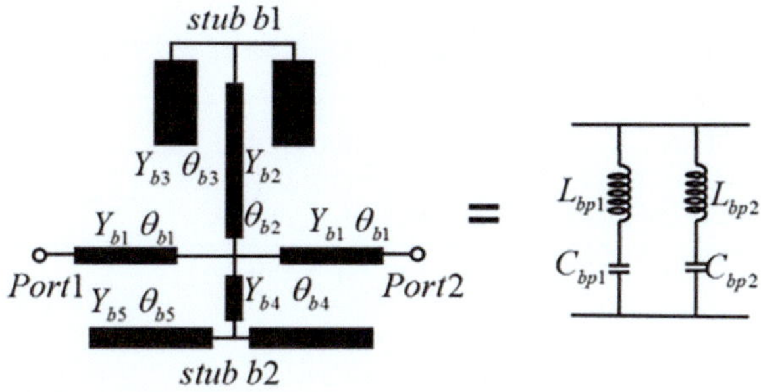

Fig. 5.3 Layout of proposed dual-band bandstop unit and its equivalent circuit

It can be inferred from Eq. (5.8) that there will be two center frequencies under different values of the two loaded stubs. The first and second stopbands are generated by stub b1 and stub b2, respectively. Also, the two center frequencies of the two stopbands are independent of the transmission lines (Y_{b1}, θ_{b1}) and mainly rely on the admittance ratios R_{b1}, R_{b2} and the physical length of the loaded stubs (l_{b2}, l_{b3}, l_{b4} and l_{b5}). Compared with the bandstop unit presented in Fig. 5.1, there is an additional bandstop band generated by stub b2. Furthermore, when R_{b1} and R_{b2} keep constant, the frequencies can be freely tuned against the physical lengths of the stubs. According to Eq. (5.8), taking l_{b3} and l_{b5} for examples, the simulated results are shown in Fig. 5.4. The two stopbands of the proposed unit can be independently tuned. The first stopband moves to the lower frequency domain as l_{b3} increases from 5.0 to 7.3 mm while the second stopband keeping constant. During the tuning, l_{b2}, R_{b1} and R_{b2} keep constant as $l_{b2} = 12.8$ mm, $R_{b1} = 1.8$, $R_{b2} = 0.9$. Moreover, the center frequency f_2 of the second stopband can also be tuned by l_{b5} without changing f_1.

Based on the same scheme of the bandwidth presented in Eqs. (5.4)–(5.7), we can achieve the bandwidth equations of the dual-band stopband unit as

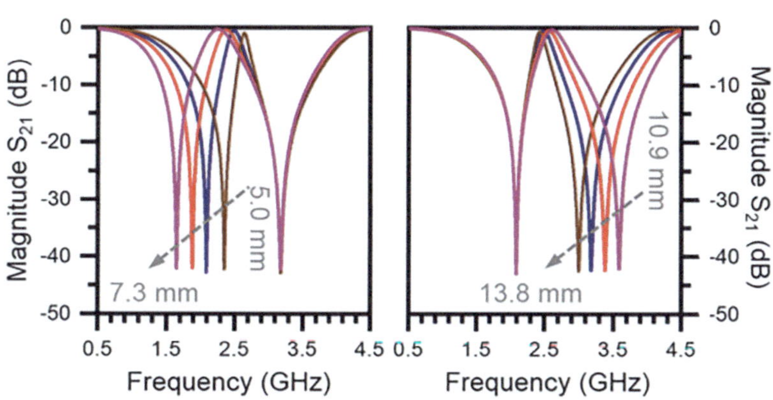

Fig. 5.4 Tuning of center frequencies of the dual-band stopband unit

$$\frac{\omega_i}{2} \cdot \frac{dX_{loadbi}}{d\omega}\Big|_{\omega=\omega_i} = \frac{1}{g_1 \Delta_i}, \quad i = 1 \text{ and } 2 \tag{5.9}$$

Then, X_{loadbi} can be obtained by the impedance-admittance transformation as

$$X_{loadb1} = -\frac{1}{Y_{b2}} \cdot \frac{1 - 2R_{b1} \tan\theta_{b3} \tan\theta_{b2}}{(2R_{b1} \tan\theta_{b3} + \tan\theta_{b2})}, \quad X_{loadb2} = -\frac{1}{Y_{b4}} \cdot \frac{1 - 2R_{b2} \tan\theta_{b5} \tan\theta_{b4}}{(2R_{b2} \tan\theta_{b5} + \tan\theta_{b4})} \tag{5.10}$$

As depicted by Eqs. (5.9)–(5.10), the two bandwidths of the stopbands can be independently controlled by (R_{b1}, Y_{b2}) and (R_{b2}, Y_{b4}), respectively, while keeping the physical lengths of the stubs constant. Figure 5.5 shows the variation of the bandwidths of the bandstop unit versus various impedance ratios (R_{b1}, R_{b2}), Y_{b2} and Y_{b4}. An observation can be made here that the bandwidths can be completely tuned by properly choosing (R_{b1}, Y_{b2}) and (R_{b2}, Y_{b4}), without changing the center frequencies. In the following part, the first stopband is taken to illustrate how to achieve this property. As presented in Eqs. (5.9) and (5.10), the bandwidth can be enlarged by increasing the values of Y_{b2}. Then, in order to keep the center frequency fixed, R_{b1} is tuned to a smaller value as shown by Eq. (5.10). Moreover, the bandwidth of the first stopband can be enlarged without changing f_1. The method is also suitable to the second stopband. Moreover, R_{b1}, Y_{b2} and R_{b2}, Y_{b4} are uncorrelated so that the bandwidths of the two stopbands can be independently controlled. These unique properties shown in Figs. 5.4 and 5.5 can be adopted to design DBBSFs with controllable bandwidths and center frequencies.

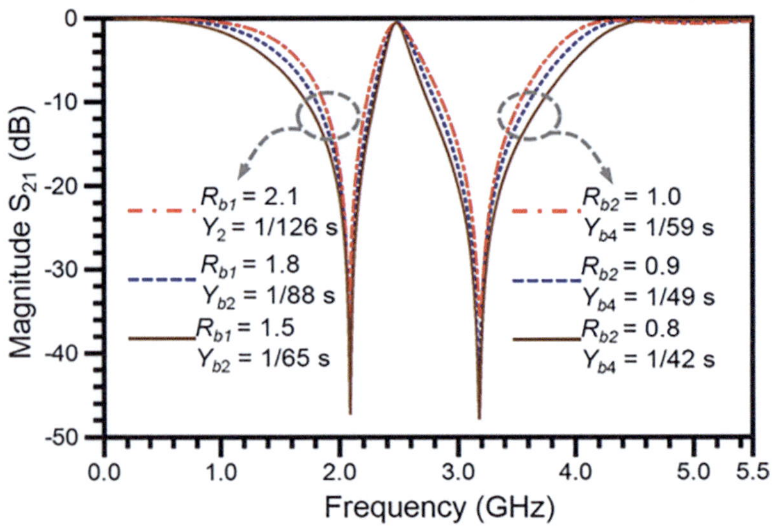

Fig. 5.5 Bandwidths variation against admittance ratios and Y_{b2}, Y_{b4} with constant $l_{b2} = 10$ mm, $l_{b3} = 5.0$ mm, $l_{b4} = 2.6$ mm, $l_{b5} = 12.8$ mm

5.2.3 *Tunable Stopband Resonant Unit*

The configuration of the proposed tunable bandstop filter unit consists of two transmission lines (characteristic admittance Y_1, electrical length θ_1) and a center loaded stepped-impedance stub (characteristic admittance Y_2, Y_3 and electrical length θ_2, θ_3) with a tunable capacitor C_L, as illustrated in Fig. 5.6. Obviously, the structure of the proposed bandstop unit is very simple so that the design complexity can be significantly reduced. In addition, the characteristic impedances of the microstrip lines at the input/output ports are $Z_0 = 1/Y_0 = 50\ \Omega$. The theory and method to independently control the tunable stopband unit will be presented in the following section [24].

Tuning of Center Frequency

The *ABCD* matrices of the transmission lines, loaded stub (A_{line1}-transmission line1, A_{stub}-loaded stub) are

$$A_{line1} = \begin{bmatrix} \cos\theta_1 & j\sin\theta_1/Y_1 \\ jY_1\sin\theta_1 & \cos\theta_1 \end{bmatrix} \tag{5.11}$$

$$A_{stub} = \begin{bmatrix} 1 & 0 \\ Y_{load} & 1 \end{bmatrix} \tag{5.12}$$

where

$$Y_{load} = jY_2 \cdot \frac{2Y_3\tan\theta_3 + \omega C_L + Y_2\tan\theta_2}{Y_2 - (2Y_3\tan\theta_3 + \omega C_L)\tan\theta_2}.$$

Then, the overall *A* matrix of the tunable bandstop unit are

$$\begin{bmatrix} A_s & B_s \\ C_s & D_s \end{bmatrix} = A_{line1} \times A_{stub} \times A_{line1} \tag{5.13}$$

where

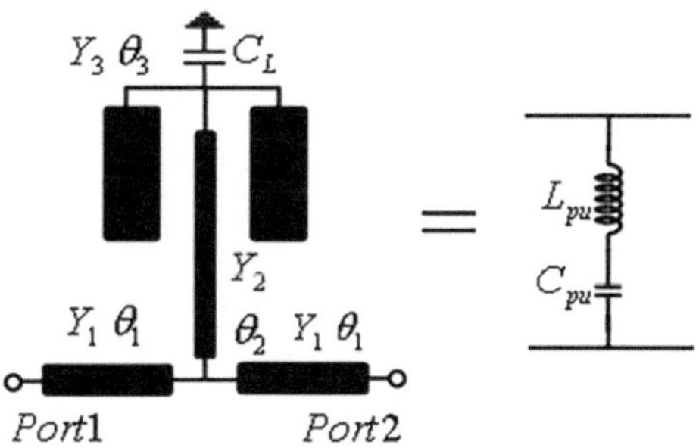

Fig. 5.6 Layout of the proposed bandstop unit and its equivalent circuit

$$A_s = D_s = \cos 2\theta_1 + jY_{load} \sin 2\theta_1/2Y_1,$$

$$B_s = -Y_{load} \sin^2\theta_1/Y_1^2 + j\sin 2\theta_1/Y_1, \;\; C_s = Y_{load}\cos^2\theta_1 + jY_1 \sin 2\theta_1.$$

The insertion loss of ABCD matrix is given by [22]

$$S_{21} = \frac{2}{A_s + B_s/Z_0 + C_s Z_0 + D_s} \tag{5.14}$$

To suppress desired frequency bands, the insertion loss must be equal to zero at the center frequency of the stopband. By equating $|S_{21}| = 0$, it yields

$$Y_2 - (2Y_3 \tan\theta_3 + \omega C_L)\tan\theta_2 = 0 \tag{5.15}$$

Equation (5.15) shows that the center frequency of the stopband is independent of the transmission lines (Y_1, θ_1) and mainly relies on the parameters of the loaded stub. For design convenience, Y_1 is chosen to be Y_0. Let $\theta = \beta l$, $\omega = 2\pi f$ and $R = Y_3/Y_2$. Then, Eq. (5.15) can be rewritten as

$$\frac{1}{\tan\left(\frac{2\pi f\sqrt{\varepsilon_r} l_2}{c}\right)} = \left(2R\tan\left(\frac{2\pi f\sqrt{\varepsilon_r} l_3}{c}\right) + \frac{2\pi f C_L}{Y_2}\right) \tag{5.16}$$

where c and ε_r are the speed of the light in free space and permittivity of a dielectric substrate, respectively. l_n is the corresponding physical length of θ_n ($n = 1$, 2 and 3). According to Eq. (5.16), the tuning range of the center frequency of the tunable bandstop unit can be determined by properly choosing l_2, l_3, Y_2, R and C_L.

Bandwidth Control

Figure 5.6 also shows the equivalent circuit [25] of the tunable bandstop unit. The stub with a loaded varactor is equivalent to the series connected inductor L_{pu} and capacitor C_{pu}. The frequency mapping from low-pass to the bandstop yields the *LC* elements as

$$L_{pu} = \frac{1}{\omega_{0u} g_{1u} \Delta_u}, \quad C_{pu} = \frac{g_{1u}\Delta_u}{\omega_{0u}} \tag{5.17}$$

$$\Delta_u = \frac{2\pi ABW_u}{\omega_{0u}} \tag{5.18}$$

where Δ_u is the fractional bandwidth of the tunable bandstop unit, ABW_u is the absolute bandwidth of the single bandstop unit, ω_{0u} is the center angular frequency of the bandstop unit, g_{1u} is the element value of the low-pass prototype mapping to the bandstop unit.

Note that, as depicted before, the center frequency of the stopband completely relies on the parameters of the loaded stub. Therefore, the loaded stub and its equivalent circuit must exhibit the same reactance slope b to obtain the required bandwidth. Thus, the reactance slope parameter can be applied for this bandwidth analysis [23]

$$b = \frac{\omega_0}{2} \cdot \frac{dX}{d\omega}\Big|_{\omega=\omega_0} \tag{5.19}$$

By equating the reactance slope parameters of the loaded stub and equivalent circuit, the following equation can be obtained to calculate the bandwidth

$$\frac{dX_{load}}{d\omega}\Big|_{\omega=\omega_{0u}} = \frac{1}{\pi g_1 ABW_u} \tag{5.20}$$

where X_{load} can be obtained by the impedance-admittance transformation $Z_{load} = 1/Y_{load}$ as

$$X_{load} = \frac{1}{Y_2} \cdot \frac{(2R\tan\theta_3 + \omega C_L)\tan\theta_2 - 1}{\left(2R\tan\theta_3 + \frac{\omega C_L}{Y_2} + \tan\theta_2\right)} \tag{5.21}$$

As explained above, with Eqs. (5.16) and (5.20), the frequency range and bandwidth can be calculated by the parameters of the stub. In order to ulteriorly confirm the scheme of the bandwidth over the tuning range, Fig. 5.7 plots the calculated results by using Eqs. (5.16) and (5.20), taking 3-dB absolute bandwidth as an example. For a Chebyshev lowpass filter with a passband ripple $L_{Ar} = 0.1$ dB, g_{1u} is selected to be 0.3052. As depicted in Fig. 5.7a, l_2 and l_3 only influence tuning range while the 3-dB absolute bandwidth almost keeps unchanged. As l_2 or l_3 increases, the whole tuning range moves down to the lower frequency band. Bandwidth mainly depends on the value of Y_2 and slightly influenced by R. The calculated results against Y_2 are shown in Fig. 5.8b. While keeping l_2, l_3 and R constant, tuning range and bandwidth can be controlled by Y_2. The higher Y_2 is, the larger bandwidth becomes. Meanwhile, through the calculation, it finds that the variation of R only introduces an influence on tuning range and the bandwidth of the unit can hardly be influenced.

A conclusion can be made here that the bandwidth within the tuning range mainly depends on Y_2. Furthermore, the frequency tuning range is decided by the Y_2, R, l_2, l_3 and C_L. Finally, a tunable bandstop unit with CABW can be achieved. Due to its simple structure and fully controlled bandstop characteristic, i.e., center frequency, absolute bandwidth, etc., the unit shows its attractions in tunable BSFs with CABW.

To verify the tunable bandstop unit configuration, a full wave electromagnetic (EM) simulation is carried out under Advanced Design Systems (ADS). Figure 5.8 shows that the simulation results coincide with the calculated results presented in Fig. 5.7. As Fig. 5.8a depicted, the bandwidth of the unit can be tuned by Y_2 and the center frequency can keep unchanged by varying R according to Eq. (5.16). Then, the parameters of the loaded stub are chosen as follows: $l_2 = 8.1$ mm, $Y_2 = 1/87.6$ s, $l_3 = 4.5$ mm, $Y_3 = 1/49.1$ s, $Y_1 = Y_0$, $l_1 = 4.7$ mm. By varying the value of C_L from

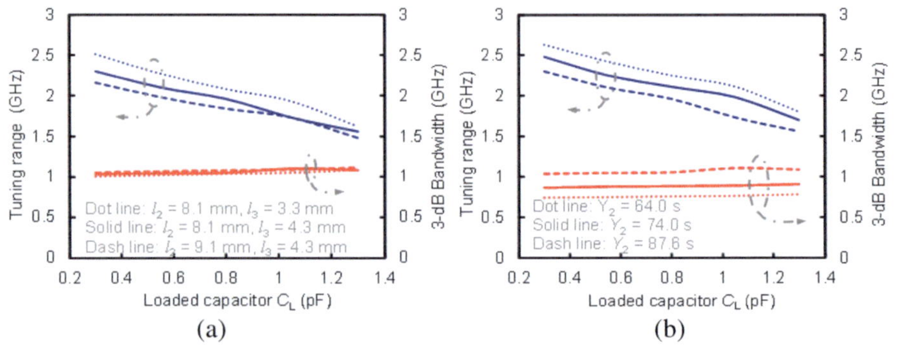

Fig. 5.7 Calculated results about the frequency tuning range and 3-dB absolute bandwidth. **a** Various l_2, l_3 with constant Y_2 and R. **b** Various Y_2 with constant l_2, l_3 and R

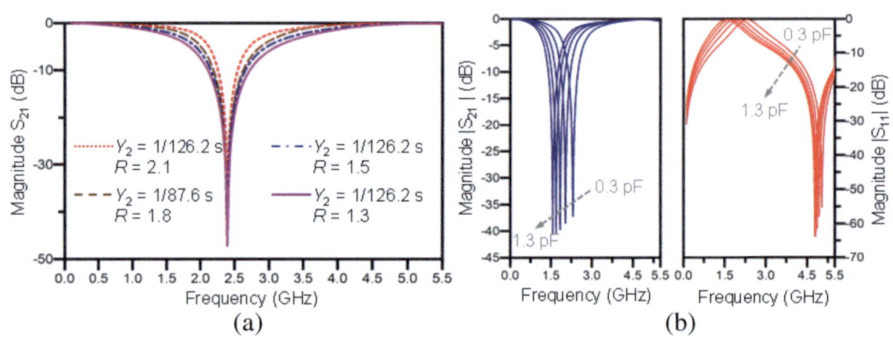

Fig. 5.8 Simulated results under ADS. **a** The bandwidth variation under Y_2. **b** S-parameters of the tunable unit against different value of the loaded C_L

1.3 to 0.3 pF, the resonant frequency increases from 1.56 to 2.30 GHz, keeping the 3 dB absolute bandwidth 1.05 ± 0.05 GHz. Therefore, a tunable bandstop unit with constant bandwidth has been achieved by the method presented in this section.

5.3 Bandstop Filter Design

Adopting the proposed bandstop units, the single-band BSFs will be addressed. Then, the bandstop unit is transformed to possess dual-stopband characteristics. Derived from the dual-stopband unit, the dual-band bandstop filter (DBBSF) is designed. Also, the design of tunable BSFs is addressed based on the basic analysis of the tunable bandstop unit. The filters are designed on a substrate with a relative dielectric constant of 2.65, a thickness of 0.5 mm and a tangent of 0.001.

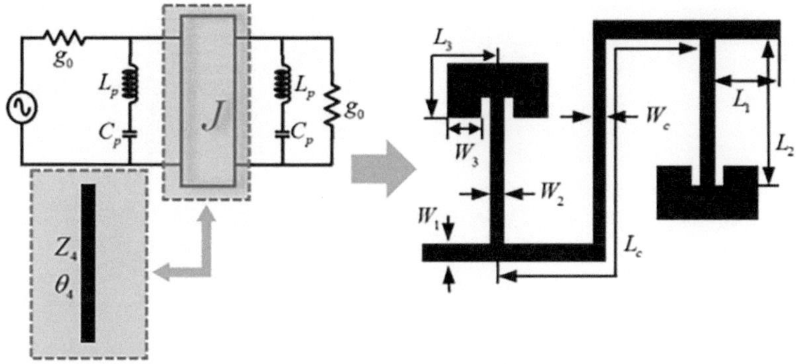

Fig. 5.9 Equivalent circuit using series *LC* resonators and schematic of 2-order BSF based on the proposed bandstop unit and J-inverter

5.3.1 Single-Band Bandstop Filter

In this section, a 2-order BSF is proposed based the above basic analysis of the bandstop unit. Figure 5.9 shows the equivalent circuit of the 2-order BSF based on the bandstop units and an ideal J inverter, which is realized by a quarter-wave transmission line [26]. The values of J-inverter are mainly controlled by the impedance Z_4 while its overall physical length is supposed to be the quarter wavelength at the center frequency. Then, the two bandstop units are arranged centre-symmetrical. Consequently, we can achieve the whole structure of the proposed BSF, as depicted on the right-side of Fig. 5.9. This structure reduces the circuit size while keeping the filter performance and designing high-order BSFs. Finally, a 2-order BSF with independently controlled bandwidth and center frequency can be achieved.

5.3.2 Dual-Band Bandstop Filter

Then, the J-inverter of the DBBSF can also be realized by adopting the same J-inverter presented in Fig. 5.10. As the value of the J-inverter is insensitive to the variation of the frequency by calculation, the values of J-inverter are supposed to be the same at the center frequencies of the two stopbands. The stubs are folded and the two bandstop units are arranged centre-symmetrical to reduce size. Finally, a compact two-order dual-band BSF with independently controlled bandwidth and stopbands can be achieved.

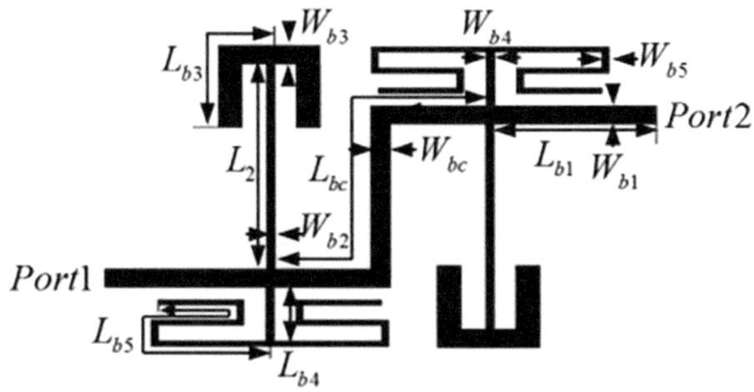

Fig. 5.10 Schematic of 2-order DBBSF based on the proposed dual-band bandstop unit

5.3.3 *Tunable Bandstop Filter*

Figure 5.11 shows the equivalent circuit of the 2-order tunable bandstop filter based on the tunable bandstop unit. The stub with a loaded varactor is equivalent to the series connected inductor L_p and a capacitor C_p here. The equation of bandwidth can be obtained as [22].

$$ABW = \frac{\omega_0^2 C_p}{2\pi J\sqrt{g_1 g_2}} \tag{5.22}$$

$$\omega_0^2 C_p \approx \frac{1}{L_{pu}} = \omega_{0u} g_1 \Delta_u = constant \tag{5.23}$$

where ω_0 is the midband angular frequency of the 2-order bandstop filter, Δ is the absolute bandwidth of the 2-order design, and g_1, g_2 is the element values of a Chebyshev prototype lowpass filter.

As discussed in Sect. 5.2.3, the tunable bandstop unit shows the unique property that the absolute bandwidth ABW_u keeps constant within the tuning range. Therefore, the relation between J-inverter and bandwidth is clear by assuming $\omega_0 \approx \omega_{0u}$ as presented by Eq. (5.23). For the 2-order tunable bandstop filter with constant absolute bandwidth, the values of the J-inverters should keep constant during the tuning range. This is the requirement of the tunable BSFs with CABW based on the bandstop units.

To obtain the desirable constant values of J-inverter, the structure of two open stubs with a transmission line is adopted here, as shown in Fig. 5.12. Since the inverter

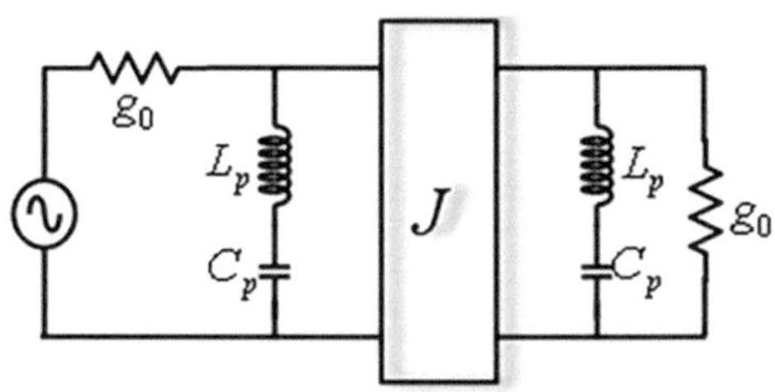

Fig. 5.11 Equivalent circuit of the tunable BSF with two bandstop units

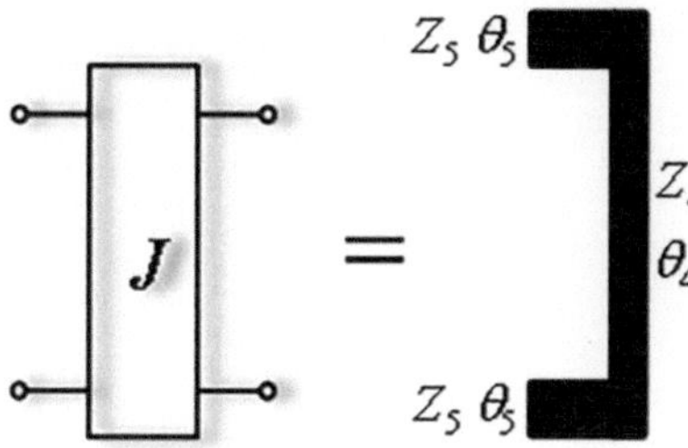

Fig. 5.12 Configuration of admittance inverter

is configured as a symmetric circuit, only the parameters *A* and *B* of the *ABCD* matrix of the inverter are needed to define the circuit. The parameters *A* and *B* are obtained as

$$A = \cos\theta_4 - \left(\frac{Z_4}{Z_5}\right)\tan\theta_5 \sin\theta_4 \tag{5.24}$$

$$B = jZ_4 \sin\theta_4 \tag{5.25}$$

For an ideal *J*-inverter, the *ABCD* matrix is expressed by

$$\begin{bmatrix} 0 & \frac{j}{J} \\ jJ & 0 \end{bmatrix} \tag{5.26}$$

Let the two *ABCD* matrixes equal, the equations for solving the *J*-inverter are given as follows:

$$Z_4 = \frac{1}{J\sin\theta_4} \tag{5.27}$$

$$Z_5 = \frac{\tan\theta_4}{J\sin\theta_4} \tag{5.28}$$

$$\theta_4 = \theta_5 = \frac{n\pi}{2}, \quad n = 1, 2, 3\ldots \tag{5.29}$$

Since the open stubs of the inverter have a much smaller input admittance than the loaded stub, they behave as an open circuit and have little influence on the response around the center frequency. In practice, the open stubs of the *J*-inverter can be neglected to save space, resulting in a decrease of bandwidth. From Eqs. (5.27) and (5.29), the values of *J*-inverter are mainly controlled by the impedance Z_4 while its overall physical length is supposed to be the quarter wavelength at the middle of the tuning frequency, just as shown in Fig. 5.13. The values of the *J*-inverter can be freely tuned against Z_4. Furthermore, the values of the *J*-inverter show a small variation during the frequency tuning range which can be regarded as constant. Therefore, the requirements of constant absolute bandwidth about *J*-inverter can be satisfied.

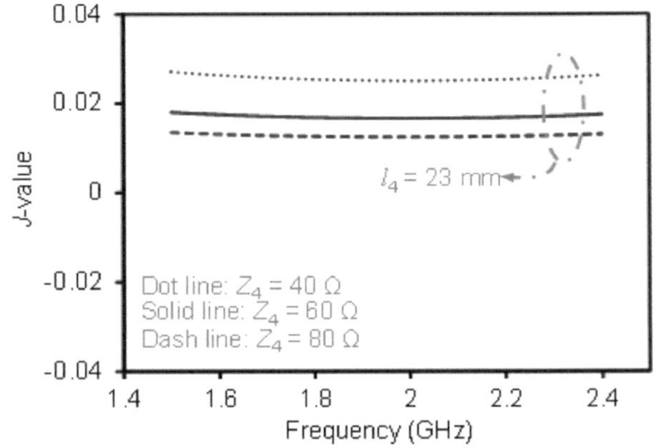

Fig. 5.13 Calculated values of admittance inverter under different Z_4 with constant l_4

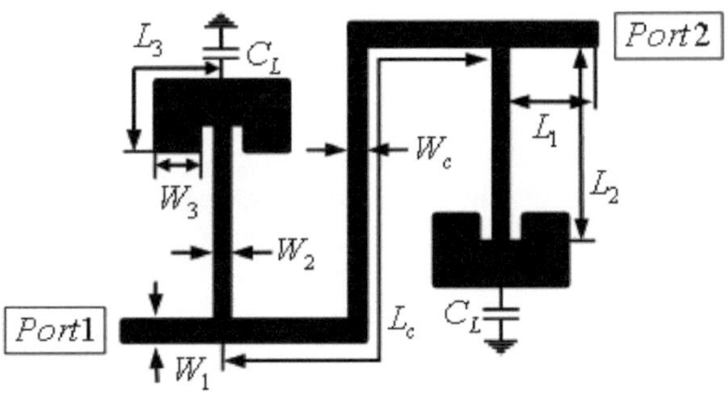

Fig. 5.14 Configuration of the 2-order tunable BSF

Finally, the two bandstop units are arranged centre-symmetrical. Meanwhile, the transmission line between them acts as a *J*-inverter, as shown in Fig. 5.14. This kind of cascade structure reduces circuit size while keeping the filter performance.

Combined with the above analysis, the design procedure of the proposed tunable bandstop filter can be summarized as follows. Step (1) *Determine* Y_2 *and the parameters of J-inverter*: Based on the requirements of CABW within the tuning range, Y_2 and the parameters of J-inverter should be optimized by properly choosing W_2, W_c and L_c. Step (2) *Determine the frequency tuning range*: According to the analysis, the tuning frequency range can be achieved by properly choosing L_2, L_3, R and variable capacitance range of varactor diodes. Step (3) *Optimize the final structure parameters*: Finally, optimize the filter performance based on full-wave and circuit simulation tools. Included with all the unexpected corner effects, discontinuity effects and via-hole effects, the ultimate circuit dimensions are settled down by full-wave simulation. The design procedure is quite easy to follow, resulting in a short reproduction time.

5.4 Validation of Proposed Bandstop Filters

To demonstrate the above analysis and formulate the design procedure, all the proposed bandstop filters are fabricated and measured. More details are listed in this section.

5.4.1 *Single-Band Bandstop Filter*

2-Order BSF

The 2-order BSF is designed with $f_0 = 2.4$ GHz, $-$ 30 dB absolute bandwidth ABW $= 320$ MHz. To meet the stopband location, $L_2 = 8.1$, $L_3 = 5.5$ (Unit: mm). The impedance ratio, Y_{t2} and J-inverter are chosen as $R = 1.8$, $Y_{t2} = 1/108$ s and J $= 0.0165$ s, respectively. The corresponding dimensions are chosen to satisfy the requirements as follows: $W_1 = 1.4$, $W_2 = 0.5$, $W_3 = 1.4$, $W_c = 1.0$, $L_c = 22.6$, L_1 $= 5.0$ (Unit: mm).

Figure 5.15 shows that the simulated and measured S-parameters agree very well with each other. The measured mid-stopband frequency is 2.36 GHz with a respective 30-dB absolute bandwidth of 300 MHz. The maximum rejection level of the stopband is around 50 dB. The insertion loss of the passband is less than 1.5 dB, which is acceptable for fabrication with such a lossy substrate. Due to the Stepped-Impedance-Resonator loaded stubs and the compact J-inverter, the overall size of the fabricated filter is only 0.21 $\lambda_0 \times$ 0.17 λ_0, which is smaller than the designs based on the traditional quarter-wavelength stubs, where λ_0 is the guide wavelength at the center frequency of the first stopband in the substrate.

3-Order BSF

In order to achieve higher attenuation levels at the midstopband, we also design a 3-order BSF referring to the same design method. The 3-order BSF is designed with $f_0 = 2.31$ GHz, $-$ 30 dB absolute bandwidth $ABW = 600$ MHz. As depicted in Fig. 5.16, the measured mid-stopband frequency of the 3-order BSF is 2.39 GHz with respective 30-dB absolute bandwidths of 580 MHz. The maximum rejection levels of the stopband are around 70 dB. The insertion loss of the passband is less than 2.5 dB.

5.4.2 *Dual-Band Bandstop Filter*

The DBBSF is designed with $f_1 = 2.1$ GHz, $f_2 = 3.36$ GHz, with $-$ 40 dB absolute bandwidth $ABW_1 = 120$ MHz, $ABW_2 = 300$ MHz. To meet the stopband locations, $L_{b2} = 10.0$, $L_{b3} = 5.5$, $L_{b4} = 2.6$, $L_{b5} = 12.4$ (Unit: mm). The impedance ratios and J-inverter are chosen as $R_{b1} = 1.8$, $R_{b2} = 0.9$, $Y_{b2} = 1/108$ s, $Y_{b4} = 1/108$ s and

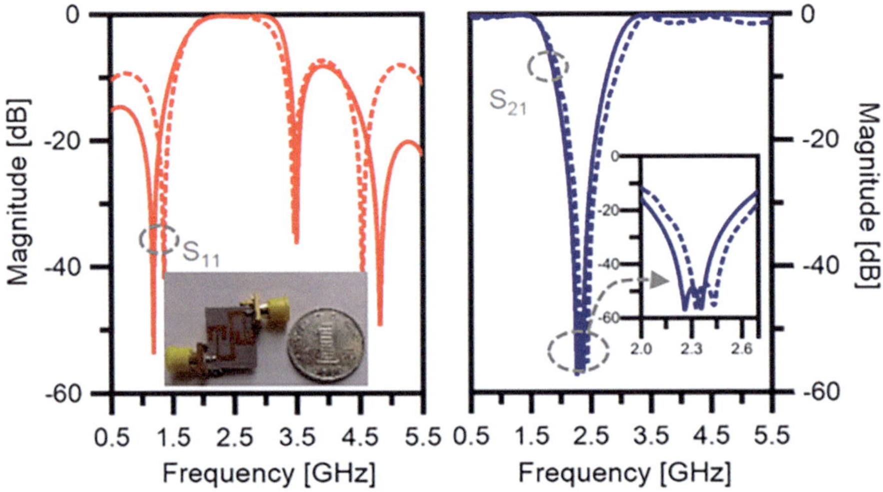

Fig. 5.15 Photograph, simulated and measured results of 2-order BSF (solid line: simulation, dot line: measurement)

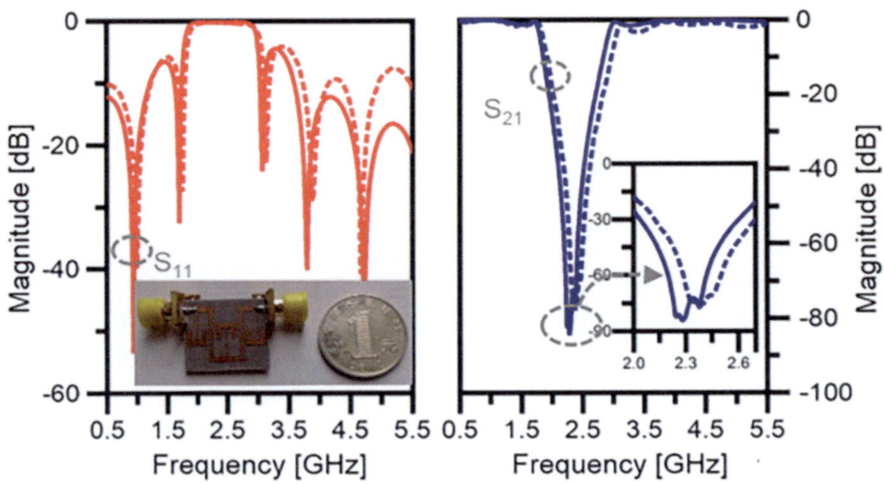

Fig. 5.16 Photograph, simulated and measured results of 3-order BSF (solid line: simulation, dot line: measurement)

J = 0.0145 s, respectively. The corresponding dimensions are chosen to satisfy the requirements as follows: $W_{b1} = 1.4$, $W_{b2} = 0.3$, $W_{b3} = 1.0$, $W_{b4} = 0.3$, $W_{b5} = 0.2$, $W_{bc} = 0.8$, $L_{bc} = 16.6$, $L_{b1} = 9.2$ (Unit: mm). The photo, simulated and measured results of the fabricated filter are presented in Fig. 5.17. The two mid-stopband frequencies are 2.14 and 3.4 GHz with respective 40-dB absolute bandwidths of 115 and 296 MHz. The maximum rejection levels of the stopbands are both around

60 dB. The overall size of the fabricated filter is only 0.21 $\lambda_0 \times 0.17\ \lambda_0$, which is also smaller than the one of the filter based on the traditional quarter-wavelength stubs, where λ_0 is the guide wavelength at the center frequency of the first stopband in the substrate.

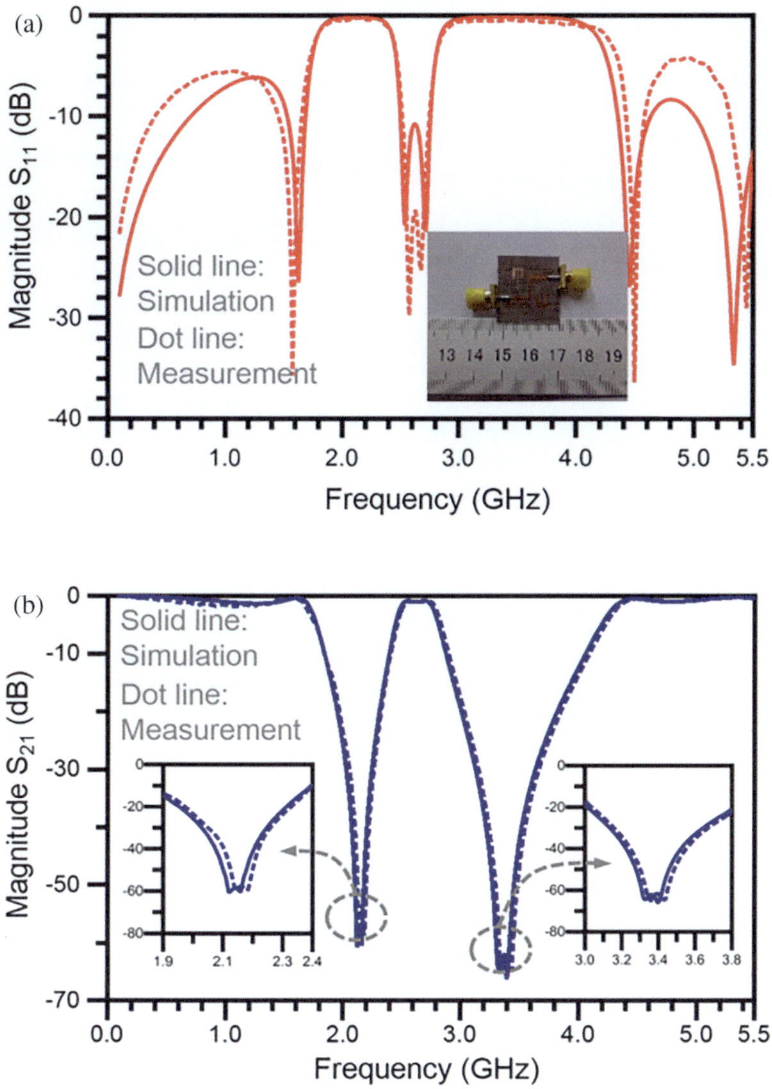

Fig. 5.17 Photograph, simulated and measured results of 2-order DBBSF (solid line: simulation, dot line: measurement). **a** S_{11} and photograph of the filter. **b** S_{21}

5.4.3 Tunable Bandstop Filter

Finally, to demonstrate this proposed analysis and design procedure of the tunable BSF experimentally, the 2-order and 3-order tunable bandstop filters are fabricated and measured.

2-Order Tunable BSF

The 2-order tunable bandstop filter is designed to have the following specifications:

35-dB stopband bandwidth: 200 MHz
Tuning Range: 1.55–2.21 GHz covers the mobile communication band around 2 GHz
Orders: 2.

To achieve the desired values of J-inverter in Fig. 5.12, the width of W_c and length of L_c are chosen according to Eqs. (5.27) and (5.29) so that the values of the J-inverter could satisfy the desired values over the tuning range. Also, according to the bandwidth design goal, Y_2 is selected as 0.0114 s and the parameters of J are $W_c = 1.0$ mm, $L_c = 23$ mm. The calculated results can meet the requirement of J-inverter, as depicted by the dot line in Fig. 5.18. To derive the required design parameters of the tuning range, L_2 and L_3 are chosen as 8.1 mm and 4.4 mm, respectively. The terminal loaded capacitors C_L were realized by connecting a SKYWORKS SMV-1405 with SC-70 package. The single varactor capacitance is 0.63 pF and 2.67 pF at 30 V and 0 V reverse bias, respectively.

The photo of the fabricated tunable bandstop filter is presented in Fig. 5.19. The dimensions of the proposed tunable BSF are as follows: $L_1 = 4.7$ mm, $L_2 = 8.1$ mm, $L_3 = 4.4$ mm, $L_4 = 3.3$ mm, $L_c = 23$ mm, $W_1 = 1.4$ mm, $W_2 = 0.5$ mm, $W_3 = 1.4$ mm, $W_c = 1.0$ mm. Only one type of DC voltage $V1$ is needed to tune the center frequency. The DC biasing is done by using a 50 K resistor. The S-parameters of the filter were measured with an Agilent E5071C vector network analyzer.

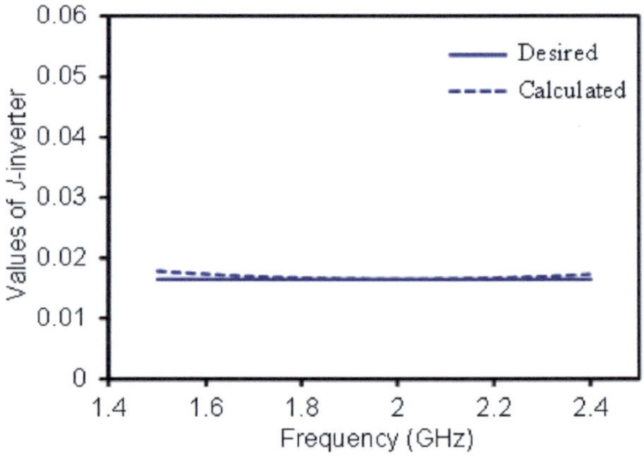

Fig. 5.18 Desired and calculated coupling coefficient

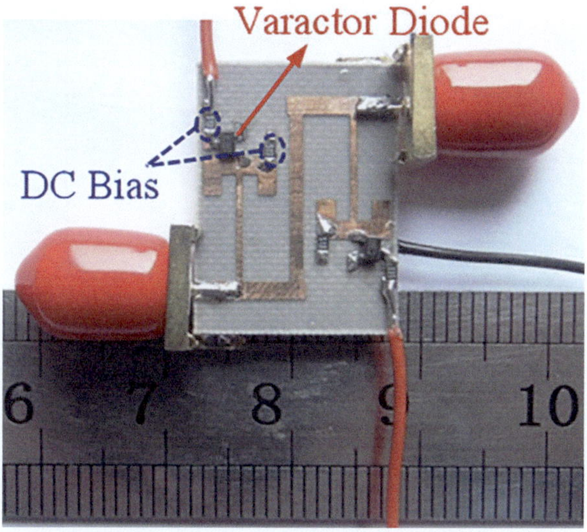

Fig. 5.19 Photograph of the fabricated two-order tunable BSF

Figure 5.20 shows the simulated and measured *S*-parameters, which agree very well with each other. As shown in Fig. 5.20a, the center frequency of the stop-band can be continuously tuned from 1.55 to 2.21 GHz and a 43% fractional tuning range can be attained. The maximum rejection levels at the midstopband are around 50 dB. Moreover, the maximum rejection levels almost keep constant during the tuning range. As the frequency tuned, the 35-dB absolutes bandwidth with a various frequency is 200 ± 25 MHz, which can be taken as constant. Figure 5.20b shows the wideband response corresponding to the bias voltage $V1$. The passband insertion loss is around 0.7 dB.

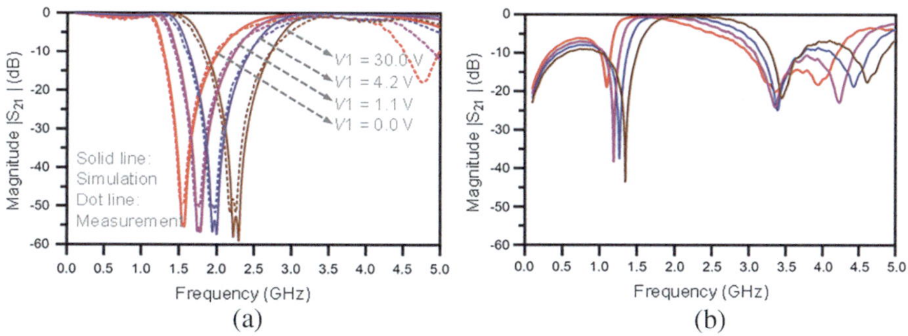

Fig. 5.20 Simulated and measured results. **a** S_{21} of the 2-order tunable BSF under different bias voltage $V1$. **b** Wideband response under the bias voltage $V1$

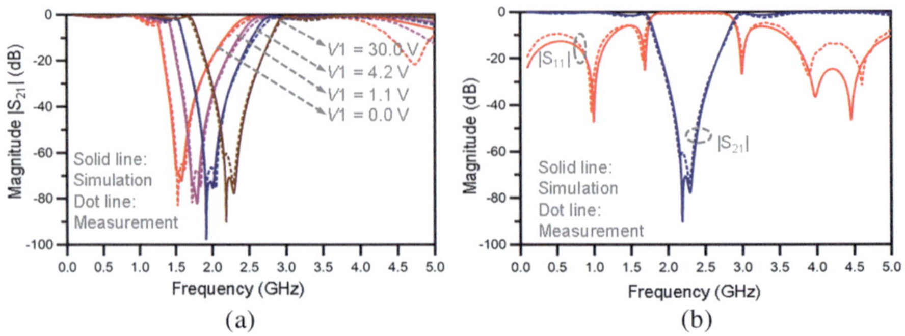

Fig. 5.21 Simulated and measured results. **a** S_{21} of the 3-order tunable BSF under different bias voltage $V1$. **b** Wideband response with the bias at 30 V

3-Order Tunable BSF

In order to achieve higher attenuation levels at the midstopband, we also design a 3-order tunable filter with CABW referring to the design procedure. The 3-order tunable bandstop filter is designed to with a 40-dB absolute bandwidth of 325 MHz. The frequency tuning range also covers the same mobile communication band around 2 GHz between 1.55 and 2.21 GHz. The cascade parts are the same as stated previously. Figure 5.21 shows the simulated and measured S_{21} which agree very well with each other. As shown in Fig. 5.21a, the center frequency of the stopband can also be continuously tuned from 1.55 to 2.21 GHz. The attenuation levels at the mid stopband are around 70 dB and keep almost unchanged while tuning the center frequency. As the frequency is tuned, the $-$ 40 dB absolutes bandwidth with a various frequency is 325 $\pm$ 35 MHz, which can be taken as constant. Figure 5.21b shows the wideband response when the bias voltage is 30 V. The passband insertion loss is around 1 dB, and the return loss is greater than 6 dB.

5.5 Conclusion

In this chapter, novel bandstop filters with independently controlled bandwidth and center frequency are proposed. The design equations have been presented. 2-order, 3-order and dual-band BSF are designed to verify the proposed design method, while good agreement between simulation and measurement is obtained. The independently controllable bandwidth and frequency, high rejection levels at midstopbands and compactness make the proposed BSFs attractive to modern communication systems. Then, a novel tunable bandstop unit has been investigated for tunable bandstop designs. Compact cascade structure is adopted to meet the requirement of constant absolute bandwidth while designing tunable filters. By tuning the bias voltage applied to the varactor diodes, the center frequency has a 43% tuning fractional range. The maximum rejection levels at the midstopband are around 50 dB and 70 dB in 2-order and 3-order design, respectively. These unique properties of the proposed tunable bandstop filter with constant absolute bandwidth can be well applied in multiband and wideband reconfigurable systems.

References

1. Wang, Y. Z. & Her, M. L. (2006). Compact microstrip bandstop filters using stepped-impedance resonator (SIR) and spur-line sections. *IET Proceedings of Microwaves, Antennas and Propagation*, 153, 435–440. https://doi.org/10.1049/ip-map:20050204.
2. Shaman, H. & Hong, J. S. (2007). Wideband bandstop filter with cross-coupling. *IEEE Transactions on Microwave Theory and Technology*, 55, 1780–1785. https://doi.org/10.1109/TMTT.2007.901600.
3. Her, M. L., Hsu, M. W., Ko, W., Wang, Y. L., Wang, K. M., Wang, C. A. & Hsu, C. J. (2008). Design and implementation of capacitive coupled-line structure bandstop filters. *Microwave and Optical Technology Letters*, 50, 1363–1368. https://doi.org/10.1002/mop.23374.
4. Divyabramham, K., Mandal, M. K. & Sanyal, S. (2008). Sharp-Rejection wideband bandstop filter. *IEEE Microwave and Wireless Component Letters*, 18, 662–664. https://doi.org/10.1109/LMWC.2008.2003452.
5. Mandal, M. K., Divyabramham, K., & Velidi, V. V. (2012). Compact wideband bandstop filter with five transmission zeros. *IEEE Microwave and Wireless Component Letters*, 22, 4–6. https://doi.org/10.1109/LMWC.2011.2173928.
6. Rudzinski, A., Lysiuk, A. & Godziszewski, K. (2013). Compact bandstop filter for UHF in modified meander arrangement with capacitors. *Microwave and Optical Technology Letters*, 55, 2110–2113. https://doi.org/10.1002/mop.27756.
7. Uchida, H., Kamino, H., Totani, K., Yoneda, N., Miyazaki, M., Konishi, Y., Makino, S., Hirokawa, J., & Ando, M. (2004). Dual-band-rejection filter for distortion reduction in RF transmitters. *IEEE Transactions on Microwave Theory and Technology*, 52, 1093–1096. https://doi.org/10.1109/MWSYM.2004.1339175.
8. Tseng, C. H. & Itoh, T. (2006). Dual-band bandpass and bandstop filters using composite right/left-handed metamaterial transmission lines. *IEEE MTT-S International Microwave Symposium*, 931–934. https://doi.org/10.1109/MWSYM.2006.249869.
9. Tang, W. X. & Hong, J. S. (2010). Coupled stepped-impedance-resonator bandstop filter. *IET Microwaves, Antennas and Propagation*, 4, 1283–1289. https://doi.org/10.1049/iet-map.2009.0419.
10. Hazeri, A. R. (2011). Compact dual-band open-stub bandstop filter with suppression of spurious passbands. *International Journal of Electronics*, 98, 679–683. https://doi.org/10.1080/00207217.2010.547808.
11. Velidi, V. K., Ajaybabu, A. & Sanyal, S. (2009). Compact tapped stepped impedance open stub dual-band bandstop filters with sharp rejection characteristics. *Microwave and Optical Technology Letters*, 51, 2274–2277. https://doi.org/10.1002/mop.24635.
12. Chin, K. S., Yeh, J. H. & Chao, S. H. (2007). Compact dual-band bandstop filters sing stepped-impedance resonators. *IEEE Microwave and Wireless Component Letters*, 17, 849–851. https://doi.org/10.1109/LMWC.2007.910481.
13. Verma, A. K., Abdel-Rahman, A., Kumar, A., Chaudhari, N. P., Balalem, A. & Omar, A. (2013). New compact dual-band bandstop filter. *International Journal of Electronics*, 100, 497–507. https://doi.org/10.1080/00207217.2012.713020.
14. Che, W. Q., Feng, W. J. & Deng, K. (2010). Microstrip dual-band bandstop filter of defected ground structure and stepped impedance resonators. *International Journal of Electronics*, 97, 1351–1359. https://doi.org/10.1080/00207217.2010.488907.

15. Cameron, R. J., Yu, M. & Wang, Y. (2005). Direct-coupled microwave filters with single and dual stopbands. *IEEE Transactions on Microwave Theory and Technology*, 53, 3288–3297. https://doi.org/10.1109/TMTT.2005.859032.
16. Y. H. Chun, J. S. Hong, P. Bao and T. J. Jackson, "Tunable slotted ground structured bandstop filter with BST varactors," *IET Microw. Antenna Propagat.*, vol. 3, no. 5, pp. 870–876, Aug. 2009.
17. I. Reines, S.-J. Park, and G. M. Rebeiz, "Compact low-loss tunable X-band bandstop filter with miniature RF-MEMS switches," *IEEE Trans. Microw. Theory Tech.*, vol. 58, no. 7, pp. 1887–1895, Jul. 2010.
18. D. R. Jachowski and C. Rauscher, "Frequency-agile bandstop filter with tunable attenuation," in Proc. *IEEE MTT-S Int Microw. Symp. Dig.*, 2009, pp. 649–652.
19. A. I. Abunjaileh and I. C. Hunter, "Tunable bandpass and bandstop filters based on dual-band combline structures," *IEEE Trans. Microw. Theory Tech.*, vol. 58, no. 12, pp. 3710–3719, Dec. 2010.
20. X.-Y. Zhang, C.H. Chou, Q. Xue and B.-J. Hu, "RF tunable bandstop filters with constant bandwidth based on a doublet configuration," *IEEE Trans. Ind. Electron.*, vol. 59, no. 2, pp. 1257–1265, Feb. 2012.
21. Jia, D., Feng, Q., Huang, X., & Xiang, Q. (2016). Single- and dual-band bandstop filters with fully controllable stopband characteristics. International Journal of Electronics, 103(11), 1954–1964. https://doi.org/10.1080/00207217.2016.1158325.
22. Pozar, D. M. (2005). *Microwave Engineering*, Wiley, New York.
23. Matthaei, G., Young, L. & Jones E. M. T. (1980). *Microwave Filter, Impedance-Matching Networks and Coupling Structures*, Artech House, Norwood.
24. Jia, D., Feng, Q., Huang, X., & Xiang, Q. (2014). Microstrip tunable bandstop filter with constant absolute band-width. Journal of Electromagnetic Waves and Applications, 28(7), 815–826. https://doi.org/10.1080/09205071.2014.891953.
25. Z. Ma, K. Kikuchi, Y. Kobayashi, T. Anada, and G. Hagiwara, "Novel microstrip dual-band bandstop filter with controllable dual-stopband response," *in Proc. Asia-Pacific Microw. Conf.*, pp.1177–1180, Dec. 2007.
26. Tsai, C. M., Lee, H. M. & Tsai. C. C. (2005). Planar filter design with fully controllable second passband. *IEEE Microwave and Wireless Component Letters*, 53, 3429–3439. https://doi.org/10.1109/TMTT.2005.855739.

Chapter 6
Other Tunable Filtering Advices

6.1 Introduction

In previous chapters, we analyzed the design theory and validation of tunable bandpass and bandstop filters. In this chapter, other tunable filtering devices are presented. The design theory and validation of tunable frequency-selective surfaces (FSS) are presented.

Tunable frequency-selective surfaces attracted great interest in recent years as tunable filters for multiband reflector antennas, for tunable radar absorbers, or to provide a varying transmission window in radomes [1]. Generating a dynamic frequency behavior requires the reactive characteristics of the surface to change with a tuning voltage, biasing magnetic field or changing shape. Thus, the literature concerning tunable frequency-selective surfaces can be generally classified into three categories: mechanically tunable surfaces, magnetically tunable surfaces, and electronically tunable surfaces. Mechanically tunable frequency-selective surfaces are realized by changing the shape of its elements [2–4]. The elements can be either solid-state [2] or liquid state [3, 4]. The magnetically tunable FSSs are implemented by using a ferrite substrate [5]. The permeability of the ferrite substrate can change with the external bias magnetic field. Although the magnetically and mechanically tunable FSSs do not need bias circuits, they suffer from several disadvantages, i.e. low tuning speed and small tuning range. The electronically tunable frequency-selective surfaces can be realized through conventional methods includes loading with active semiconductor devices like varactor diodes [6–10], p-i-n diode [11], or barium-strontium-titanate (BST) varactor [12] in the element design. Although the electronic tuning requires the biasing circuit for electronic devices, it has many benefits, i.e. small size, nanosecond tuning speed and low cost. Thus, the electronic tunable surfaces were widely studied in the past few years.

Recently, a novel concept of three-dimensional (3-D) frequency-selective structure (FSS) consisting of a 2-D periodic array of vertical microstrip line resonators was reported in [13–15], where multiple transmission zeros/poles can be produced

F. Nian et al., *RF & Microwave Multi-mode and Tunabe Filters*,
https://doi.org/10.1007/978-981-92-0577-6_6

by a 3-D FSS with multimode cavities/resonators. Compared with 2-D surfaces, 3-D FSSs have demonstrated greater flexibility, better filtering performance, and stable angular response.

In this chapter, a new second-order 3-D tunable bandpass FSS using varactor diodes is presented. This structure is based on 2-D periodic array of vertical varactor-loaded microstrip line resonators and a number of horizontal metallic plates. The proposed varactor-loaded $\lambda/4$ stepped-impedance resonator (SIR) is first theoretically analyzed to demonstrate its superior performance in implementation of a wide tuning range. Design parameters of the via hole and metallic plate are then properly selected in order to meet the requirement of bandwidth and location of transmission zeros. Finally, a prototype 3-D tunable bandpass FSS with 2:1 (1.4–2.75 GHz) tuning range is fabricated and measured to verify our design concept and its tuning performance. Based on this, the subsequent sections of this chapter will proceed as follows: Sect. 6.1 will give specific topology illustration of the proposed tunable FSS. Section 6.2 will provide an in-depth analysis of the operating principles of the proposed tunable FSS. Section 6.3 will introduce the validation for the proposed tunable FSS, present their simulated and measured results, summarize the performance comparisons of the proposed tunable FSS with other reported tunable bandpass FSSs. Finally, Sect. 6.4 will present a comprehensive summary of the chapter's content.

6.2 Proposed Topology of Tunable 3D FSS

The proposed tunable bandpass FSS, illustrated in Fig. 6.1a, consists of an array of printed circuit boards and a number of inserted metallic plates and varactor diodes [16]. Figure 6.1b shows the detailed biasing circuit of the individual varactor diode consisting of a RF choke and a DC blocking circuit. Figure 6.1c, d show the structural details of a unit cell, where the dash lines represent the periodic boundary. It is seen that two adjacent varactor-loaded $\lambda/4$ SIRs are coupled with each other through a via-hole. In addition, a metallic plate of thickness t is inserted in the middle of the air gap formed by the microstrip line and the upper ground, which is used to realize a bandpass filtering response by blocking the air propagation path [14]. Periods along the x- and y-axes are denoted by b and h, respectively. The thickness of proposed tunable FSS along the z-axis is represented by l. D denotes the diameter of via hole.

6.3 Theoretical Analysis

In the structure of the unit cell, two quasi-TEM modes are excited in the shielded microstrip resonator: air mode and substrate mode [14]. The substrate mode mainly contributes to the passband of the tunable FSS, whereas the air mode only influences the out-of-band rejection. Thus, the passband and stopband can be independently controlled by substrate mode and air mode, respectively. When the E-field of an

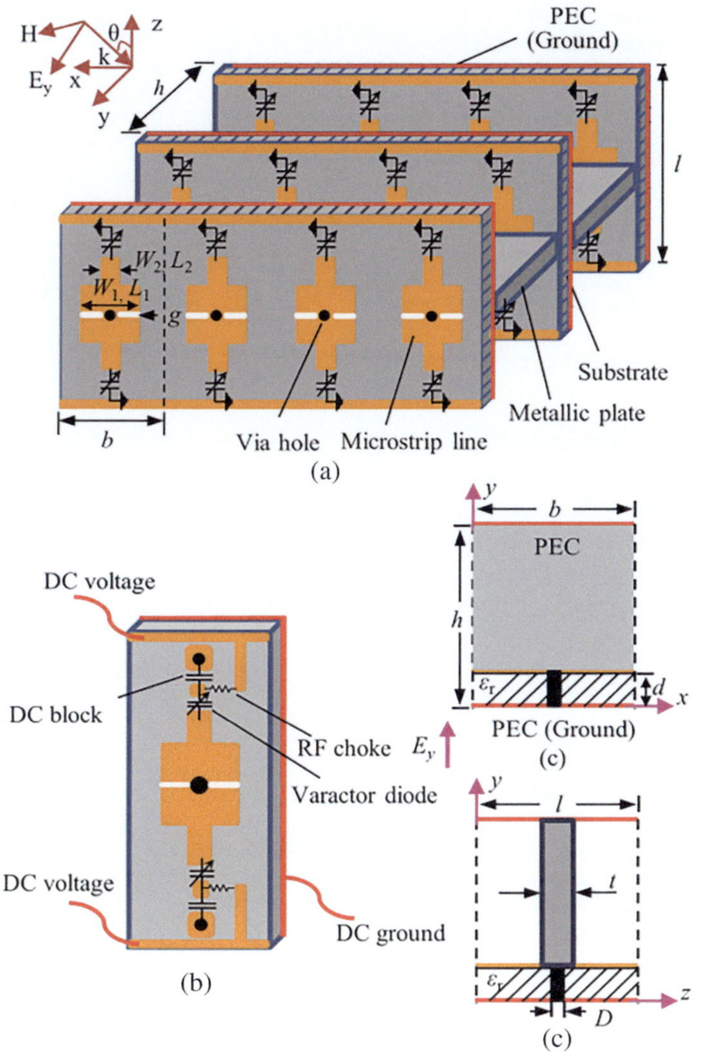

Fig. 6.1 Structure of our proposed tunable 3-D bandpass FSS. **a** Perspective view. **b** Front view of a unit cell with biasing circuit. **c** Top view of a unit cell in the xy plane. **d** Side view of a unit cell in the yz plane

incident plane wave is perpendicular to the strips (TE polarization), as indicated in Fig. 6.1, the varactor-loaded SIR-based structure will exhibit a tunable bandpass filtering performance. The tuning range of the passband controlled by the substrate mode will be discussed in the following paragraphs.

Figure 6.2 shows the schematics of the varactor-loaded λ/4 SIR, which consists of two impedance sections Z_1 and Z_2 with electrical lengths of θ_1 and θ_2, respectively, and a varactor diode is attached at the left-end of Z_2 section. The parasitic

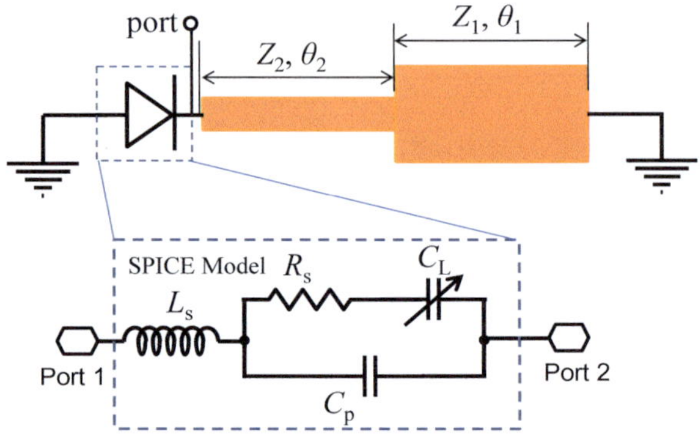

Fig. 6.2 Varactor-loaded λ/4 stepped-impedance resonator

capacitance C_P and inductance L_S in Fig. 6.2 are introduced to account for the packaging effect. Compared with junction capacitance C_L, the package capacitance C_P can be neglected. The value of R_S is small and mainly affects the insertion loss of the structure. Thus, the SPICE model of the varactor diode can be simplified as an inductance in series with a capacitance. The right-end of Z_1 section is short-circuited; the resonant frequency can thus be found from (6.1)

$$\frac{\cot\theta_1}{Z_1} = \frac{1}{Z_2}\frac{\tan\theta_2 + Z_2 Y_{load}}{1 + Z_2 Y_{load}\tan\theta_2} \tag{6.1}$$

where

$$Y_{load} \approx \frac{\omega C_L}{1 - \omega^2 L_S C_L}$$

In order to study the characteristic of this resonator, Z_1 is chosen to be 29 Ω, and the total electrical length of the transmission line ($\theta_1 + \theta_2$) is chosen to be 50°. To shift the resonant frequencies proportionally under different impedance ratios R ($R = Z_2/Z_1$) by loading the same capacitance, the tuning ranges need to be characterized. The tuning range indicates how much a resonant frequency is shifted by varying capacitance C_L. Assume that when C_L varies from C_1 to C_2, the resonant frequency shifts from f_1 and f_2. The tuning range can be defined as

$$\text{Tuning range} = 2\left|\frac{f_2 - f_1}{f_1 + f_2}\right| \times 100\% \tag{6.2}$$

Figure 6.3 shows the calculated resonant frequencies versus capacitance C_L with different values of parameters R. It indicates that reducing the impedance ratio R results in a wider tuning range, as shown in Fig. 6.3. As R varies from 2.4 to 0.4, the tuning range is changed from 45 to 66% accordingly. We can also see that the slope of three curves become lager when the value of the loading capacitance decreases, especially under a low impedance ratio R. Thus, a smaller capacitance value of

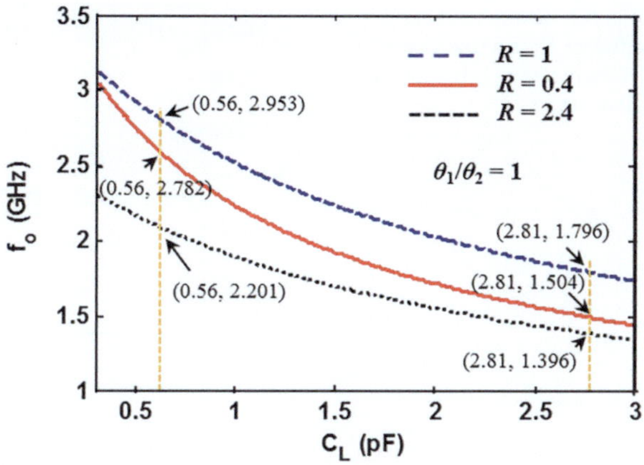

Fig. 6.3 Calculated resonant frequencies as a function of loading capacitor C_L under different values of R

varactor diode should be initially selected. The tuning range can then be extended by properly choosing the stepped-impedance resonator structure under the chosen varactor diode.

6.4 Validation of the Proposed Tunable 3D FSS

To demonstrate the above analysis and formulate the design procedure, a tunable FSS is designed using a microstrip resonator loaded with two varactor diodes to achieve the following specifications:

Tuning range: 1.40–2.75 GHz
Fractional bandwidth (FBW): 14% at 2.75 GHz
Number of Poles: 2.

To derive the required design parameters, the tunable FSS centering at 2.75 GHz has been designed initially with 3-dB fractional bandwidth of 14% based on a 0.813-mm-thick Rogers's 4003 substrate with a dielectric constant 3.55. To begin, W_2, L_2 and C_L are chosen as 4 mm, 6 mm and 0.56 pF, respectively. W_1 and L_1 can then be determined with respect to the center frequency of 2.75 GHz. Based on Fig. 6.3, C_L for realizing the tuning range of 1.40–2.75 GHz is estimated to be from 0.56 to about 2.8 pF. After that, the via hole diameter D is chosen so that the two resonators are coupled through a suitable coupling coefficient K, where K is defined as $(f_{odd} - f_{even})/f_o$, while f_{odd}, f_{even} are the resonant frequencies of odd- and even-mode equivalent circuits, as the inset figure shown in Fig. 6.4. The coupling coefficient K versus different center frequencies of tunable filter is plotted in Fig. 6.4 with three different values of diameter D. In this design, D is chosen to be 0.3 mm so that the FBW is about 14% in the 2.75 GHz. Then, h and t of the metallic plate are determined by the location of transmission zeros, which only influence the out-of-band rejection performance above the passband [14]. Finally, some fine tuning may

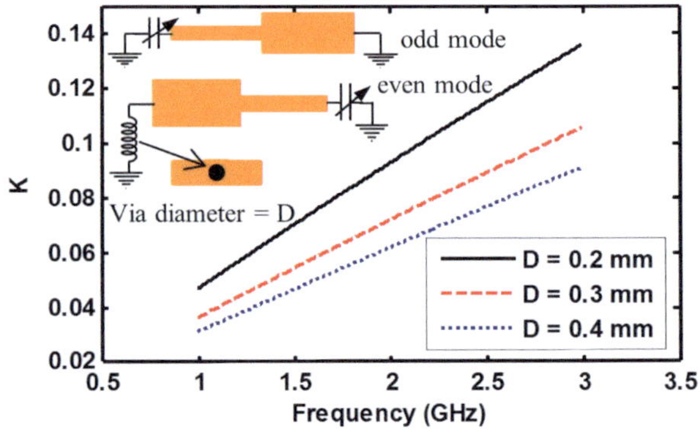

Fig. 6.4 Extracted coupling efficient K as a function of frequency with different via diameter D

be needed through a full-wave simulator, ANSYS high frequency structure simulator. The parameters for the final layout of our proposed tunable FSS are as follows: b = 6.0 mm, W_1 = 1.0 mm, L_1 = 6.0 mm, W_2 = 4.0 mm, L_2 = 6.0 mm, t = 2.4 mm, l = 32.0 mm, g = 0.2 mm, h = 4.013 mm.

The diode selected in this work is the Skyworks SMV1405-040LF, which is a silicon abrupt varactor with a nominal zero-bias junction capacitance between 0.5 and 0.6 pF. Since flexibility and low profile are desired for our design, a small package is preferred and the type with dimensions of 1.0 × 0.6 mm^2 (0402 package) is chosen. A 12-pF DC-blocking chip capacitor (ATC 600S, Rs = 0.12 Ω) is connected in series with each silicon diode to isolate the anode from the ground. The DC biasing circuit uses a 47-kΩ resistance with 0603 package to reduce the RF-signal leakage through the biasing network. Figure 6.5a shows the photography of our fabricated prototype, which is approximately 143.2 mm × 32 mm × 24.3 mm in size and consists of 22 × 6 unit cells. Figure 6.5b illustrates the assembly of several unit cells. The microstrip resonators printed on the Rogers 4003 substrate are cut into six layers of 2-D pieces. Each planar piece has a size of 143.2 mm × 32 mm, including the biasing network. It consists of 22 tunable hybrid unit cells, 44 varactor diodes, 44 DC-blocking capacitors and 44 resistors and two biasing wires. The two biasing wires connecting all unit cells are perpendicular to the incident plane wave, which will not affect the TE-polarized wave. The top metallic plate for blocking the electromagnetic wave leakage through the surface of the top circuit board is realized using aluminum due to its light weight. Six inserted copper plates (143.2 mm in length) are used to connect each of two adjacent microstrip-line circuit boards and top aluminum palate.

The fabricated tunable 3D FSS is then measured inside a parallel-plate waveguide structure, and measurement set-up has been described in [17]. The measured reflection and transmission response at normal incidence are provided in Fig. 6.6a, b. The measured return loss is higher than 10 dB in the entire tuning range. As the bias voltage varies from 0 to 30 V, the center frequency of the FSS is tuned from 1.40 to 2.75 GHz with an insertion loss in the range 4.5–1.0 dB. The overall insertion loss can be expressed by the following equation [18],

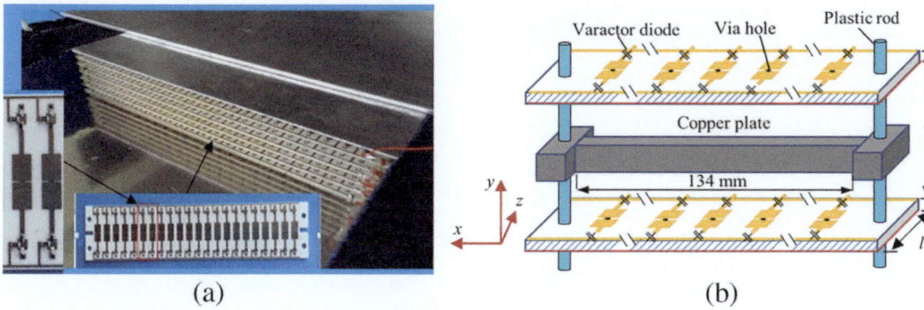

Fig. 6.5 Our designed tunable FSS. **a** Photograph. **b** Assembly details

$$IL\,(\mathrm{dB}) = 4.343\sum_{i=1}^{n}\frac{g_i}{\Delta Q_{ui}} \tag{6.3}$$

where Δ and Q_{ui} respectively represent the fractional bandwidth and the unloaded quality factor of the resonator g_{i}, which can be evaluated at the center frequency. To validate the assertion that the insertion loss is mainly dominated by the varactor diode, one example is considered here. The unloaded quality factor Q_{u} of the microstrip resonator and varactor can be calculated as [18, 19]

$$Q_M = \frac{\sqrt{L_0}}{\sqrt{L_0}-1}\left(\left|\frac{\omega_1}{\omega_0}-\frac{\omega_0}{\omega_1}\right|\right)^{-1} \tag{6.4}$$

$$Q_v = \frac{1}{2\pi f_0 C\cdot R} \tag{6.5}$$

respectively, where L_0, ω_1 and ω_0 represent the insertion loss, 3-dB angular frequency, and the center frequency of the microstrip resonator, respectively, C and R are the capacitance and parasitic resistance of the varactor, respectively. According to [19, 20], for the FSS with $f_0 = 1.4$ GHz, the Q_{u} values of the microstrip resonator and varactor diode are, 150 and 50, respectively. Therefore, the overall Q_{u} is about 37.5. Using (6.3), the simulated insertion loss of the 2-pole bandpass FSS with 9% fractional bandwidth is around 4 dB and this agrees well with the simulated insertion loss.

The simulated and measured results are also compared and shown in Fig. 6.6. The full-wave simulation results are obtained using HFSS together with lumped circuit models of varactor diodes using periodic boundaries. It is observed that simulated and measured are in good agreement in the passband. Some differences in the stopband are mainly attributed to the measurement set-up, such as the existence of an air gap between parallel-plate waveguide and the top/bottom metal layer of our proposed FSS, imperfect foam absorber, which may be improved by using the free-space method. Besides above-mentioned error sources, the inaccurate circuit model of the varactor and unwanted parasitic effects may also contribute to the difference between

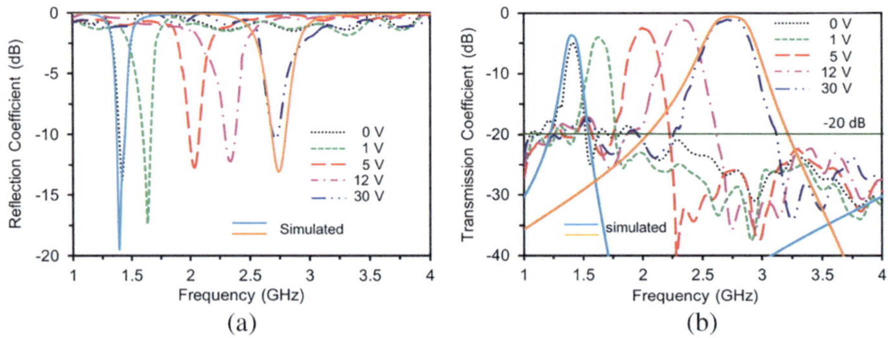

Fig. 6.6 Simulated and measured reflection and transmission coefficients of our tunable FSS for bias voltage of: 0, 1, 5, 12, 30 V (from left to right). **a** Reflection. **b** Transmission

Table 6.1 Comparison of tunable band-pass FSSs

References	Configuration	C ratio (C_{max}/C_{min})	Tuning range (%)	I.L. (dB)
[2]	3-D with mechanism	–	17	< 1.0
[3]	2-D with liquid metal	–	45	< 2.5
[4]	2-D with liquid metal	–	33	< 1.0
[9]	2-D with varactor diode	14.4	53	1.6–7.2
			58	5.4–9.5
This work	3-D with varactor diode	5	65	1.0–4.5

the measured and simulated results, including the parasitic capacitance caused by soldering and bias circuits.

Performance comparisons of the proposed tunable FSS with other reported tunable bandpass FSSs are summarized in Table 6.1. It is seen that our proposed tunable FSS can provide the largest tuning range with satisfactory insertion loss.

Although we focus on the normal incidence in this work to demonstrate the tuning performance of our 3D FSS, it is certainly important to examine the filtering performance under oblique incidences. Simulated S-parameter results of our tunable FSS for TE polarization under oblique incidence are shown in Fig. 6.7. It is observed that our proposed FSS exhibits stable frequency response under oblique incident angles up to 60°.

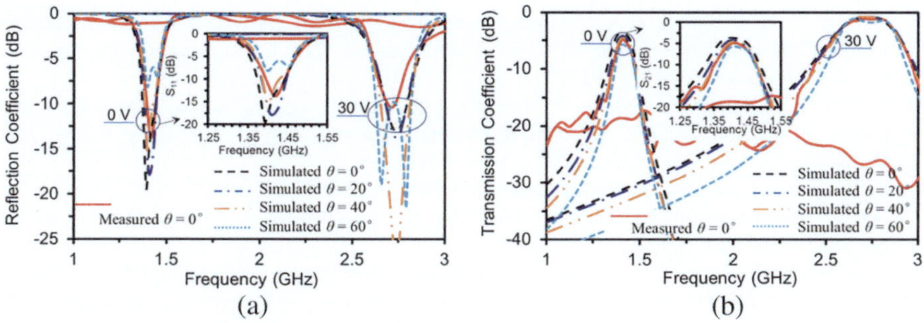

Fig. 6.7 Simulated reflection and transmission coefficients under different oblique incidence angles for TE polarization. **a** Reflection. **b** Transmission

6.5 Conclusion

This chapter has presented a novel tunable 3-D FSS with second-order band-pass responses. The structure consists of a periodic array of vertical microstrip lines loaded with lump capacitors and a number of metallic plates. The varactor-loaded $\lambda/4$ SIR for each unit cell of the proposed tunable FSS is proposed and analyzed to extend the tuning range. Based on the analysis, a second-order 3-D tunable FSS with 2:1 (1.4–2.75 GHz) tuning range has been designed, fabricated, and measured to demonstrate the tuning performance of our proposed FSS. Moreover, the tuning range can be further extended by choosing varactor diodes with a high C ratio.

References

1. L. Boccia, I. Russo, G. Amendola and G. Di Massa, "Tunable frequency-selective surfaces for beam-steering applications," Electron. Lett., vol. 45, no. 24, pp. 1213–1214, 2009.
2. S. N. Azemi, K. Ghorbani and W. S. T. Rowe, "A reconfigurable FSS using a spring resonator element," IEEE Antenna Wireless Propag. Lett., vol. 12, pp. 781–784, 2013.
3. M. Li, B. Yu and N. Behdad, "Liquid-tunable frequency selective surface," IEEE Microw. Wireless Compon. Lett., vol. 20, no. 8, pp. 423–425, Aug. 2010.
4. M. Li and N. Behdad, "Fluidically tunable frequency selectively/phase shifting surface for high-power microwave application," IEEE Trans. Antennas Propag., vol. 60, no. 6, pp. 2748–2759, June 2012.
5. T. K. Chang, R. J. Langley and E.A. Parker, "Frequency selective surfaces on biased ferrite substrate," IEE Electron. Lett., vol. 30, no. 15, pp. 1193–1194, 1994.
6. F. Bayatpur and K. Sarabandi, "A tunable metamaterial frequency-selective-surface with variable modes of operation," IEEE Trans. Antennas Propag., vol. 57, no. 6, pp. 1433–1438, June 2009.
7. C. Mias and J. H. Yap, "A varactor-tunable high impedance surface with a resistive-lumped-element biasing grid," IEEE Trans. Antennas Propag., vol. 55, no. 7, pp. 1955–1962, July 2007.

8. D. Cure, T. M. Weller and F. A. Miranda, "Study of a low-profile 2.4-GHz planar dipole antenna using a high-impedance surface with 1-D varactor tuning," IEEE Trans. Antennas Propag., vol. 61, no. 2, pp. 506–515, Feb. 2013.
9. W. Pan, C. Huang, P. Chen, M. Pu, X. Ma and X. Luo, "A beam steering horn antenna using active frequency selective surface," IEEE Trans. Antennas Propag., vol. 61, no. 12, pp. 6218–6223, Dec. 2013.
10. L. Boccia, I. Russo, G. Amendola and G. D. Massa, "Multilayer antenna-filter antenna for beam-steering transmit-array applications," IEEE Trans. Antennas Propag., vol. 60, no. 7, pp. 2287–2300, Jul. 2012.
11. B. S. Izquierdo, E. A. Parker and J. C. Batchelor, "Dual-band tunable screen using complementary split ring resonator," IEEE Trans. Antennas Propag., vol. 58, no. 11, pp. 3761–3765, Nov. 2010.
12. M. Sazegar, Y. L. Zhang, C. Kohler, H. Maune, M. Nikfalazar, J. R. Biner and R. Jakoby, "Beam steering transmit array using tunable frequency selective surface with integrated ferroelectric varactors," IEEE Trans. Antennas Propag., vol. 60, no. 12, pp. 5690–5699, Dec. 2012.
13. A. K. Rashid and Z. Shen, "A novel band-reject frequency selective surface with pseudo-elliptic response," IEEE Trans. Antennas Propag., vol. 58, no. 4, pp. 1220–1226, Apr. 2010.
14. B. Li and Z. Shen, "Three-dimensional bandpass frequency-selective structure with multiple transmission zeros," IEEE Trans. Microw. Theory Tech., vol. 61, no. 10, pp. 3578–3589, Oct. 2013.
15. B. Li and Z. Shen, "Three-dimensional dual-polarized bandpass frequency selective structures with wide out-of-band rejection," IEEE Trans. Antennas Propag., vol. 62, no. 1, pp. 130–137, Jan. 2014.
16. X. G. Huang, Z. Shen, Q. Y. Feng and B. Li, "Tunable 3-D Bandpass Frequency-Selective Structure With Wide Tuning Range," in IEEE Transactions on Antennas and Propagation, vol. 63, no. 7, pp. 3297–3301, July 2015, https://doi.org/10.1109/TAP.2015.2428737.
17. Y. Shang, Z. Shen and S. Xiao, "On the design of single-layer circuit analog absorber using double-square-loop array," IEEE Trans. Antennas Propag., vol. 61, no. 12, pp. 6022–6029, Dec. 2013.
18. J.-S. Hong and M. J. Lancaster, Microstrip Filters for RF/Microwave Applications. New York: Wiley, 2001.
19. M. Makimoto and S. Yamashita, Microwave Resonators and Filters for Wireless Communication: Theory, Design and Application, Springer, 2001.
20. SMV1405-040LF datasheet. Skyworks Solutions, Inc. 2010.

Chapter 7
Conclusion

This book focuses on RF/microwave multimode and reconfigurable filters, systematically introducing their application scenarios, main implementation methods for tunability/reconfigurability, various classifications, and typical performances of reconfigurable filters. It then elaborates, in separate chapters, on the research background, design theories, typical designs, and verifications of multimode bandpass filters, bandwidth-controllable tunable bandpass filters, multi-band tunable bandpass filters, tunable bandstop filters, and other types of tunable filtering devices. The content of each chapter can be summarized as follows:

Chapter 2 emphasizes the relevant content of multimode filters. The design of filter passbands, out-of-band rejection, and multi-passbands are achieved by controlling different modes within the resonator. Furthermore, different passband selection characteristics can be realized through mode transformation, laying a foundation for the subsequent design of tunable and reconfigurable filters.

Chapter 3 focuses on bandwidth-controllable tunable filters, with emphasis on the design methods for maintaining constant bandwidth during frequency tuning. The design and verification of 2nd–6th order bandpass filters are carried out, all of which can maintain constant bandwidth during frequency tuning. Secondly, it introduces the design method of broadband tunable bandwidth filters based on multi-mode resonators, along with example verifications.

Chapter 4 centers on multi-band tunable bandpass filters, summarizing their design methods. These methods are mainly divided into two categories: using multi-mode resonators and controlling multi-coupling channels to realize filters with independently tunable passbands. Different passband characteristics can also be achieved by controlling resonance modes and coupling properties. Subsequently, multiple specific case studies and verifications of the design methods are presented.

Chapter 5 introduces tunable bandstop filters, proposing a bandstop resonant unit that facilitates the design of multi-order and multi-stopband bandstop filters. The bandwidth variation during frequency tuning can be controlled through coupling

F. Nian et al., *RF & Microwave Multi-mode and Tunabe Filters*,
https://doi.org/10.1007/978-981-92-0577-6_7

stubs and resonant units. Case studies and verifications of 2-order, 3-order, and multi-stopband filters are conducted, along with the design and verification of tunable bandstop filters.

Chapter 6 focuses on tunable frequency selective surfaces (FSS), proposing a frequency-tunable FSS structure. Frequency tuning is achieved based on varactor diodes, and finally, case design and verification are completed.

With the rapid development of semiconductor chips, three-dimensional packaging integration, and the demand for multi-functional, integrated, and arrayed RF front-ends, tunable filtering technology is also advancing rapidly. New forms of tunable filters have emerged, such as on-chip tunable filters. These can be implemented using on-chip processes to achieve chip-level tunable filters, reducing volume and cost while being limited by the performance constraints of planar microstrip processes.

In summary, the functional requirements for tunable filters have always included a wide tuning range, low insertion loss, high out-of-band rejection, and flexible control over bandwidth, passbands, and stopbands during frequency tuning. Such tunable filters can minimize system complexity and reduce costs while meeting front-end filtering requirements. The appropriate process and type of tunable filter should be selected based on actual application scenarios.

MIX
Papier aus verantwortungsvollen Quellen
Paper from responsible sources
FSC® C105338

If you have any concerns about our products,
you can contact us on
ProductSafety@springernature.com

In case Publisher is established outside the EU,
the EU authorized representative is:
Springer Nature Customer Service Center GmbH
Europaplatz 3, 69115 Heidelberg, Germany

Printed by Libri Plureos GmbH
in Hamburg, Germany